Eureka Math
Algebra II
Modules 3 & 4

Special thanks go to the Gordon A. Cain Center and to the Department of Mathematics at Louisiana State University for their support in the development of *Eureka Math*.

This book may be purchased from the publisher at eureka-math.org
10 9 8 7 6 5 4 3

ISBN 978-1-63255-331-7

Lesson 1: Integer Exponents

Classwork

Opening Exercise

Can you fold a piece of notebook paper in half 10 times?

How thick will the folded paper be?

Will the area of the paper on the top of the folded stack be larger or smaller than a postage stamp?

Exploratory Challenge

a. What are the dimensions of your paper?

b. How thick is one sheet of paper? Explain how you decided on your answer.

c. Describe how you folded the paper.

d.   Record data in the following table based on the size and thickness of your paper.

Number of Folds	0	1	2	3	4	5	6	7	8	9	10
Thickness of the Stack (in.)											
Area of the Top of the Stack (sq. in.)											

e.   Were you able to fold a piece of notebook paper in half 10 times? Why or why not?

f.   Create a formula that approximates the height of the stack after n folds.

g.   Create a formula that will give you the approximate area of the top after n folds.

h.   Answer the original questions from the Opening Exercise. How do the actual answers compare to your original predictions?

EUREKA
MATH™

©2015 Great Minds. eureka-math.org
ALG II-M3-SE-B2-1.3.0-08.2015

Example 1: Using the Properties of Exponents to Rewrite Expressions

The table below displays the thickness and area of a folded square sheet of gold foil. In 2001, Britney Gallivan, a California high school junior, successfully folded a 100-square-inch sheet of gold foil in half 12 times to earn extra credit in her mathematics class.

Rewrite each of the table entries as a multiple of a power of 2.

Number of Folds	Thickness of the Stack (millionths of a meter)	Thickness Using a Power of 2	Area of the Top (square inches)	Area Using a Power of 2
0	0.28	$0.28 \cdot 2^0$	100	$100 \cdot 2^0$
1	0.56	$0.28 \cdot 2^1$	50	$100 \cdot 2^{-1}$
2	1.12		25	
3	2.24		12.5	
4	4.48		6.25	
5	8.96		3.125	
6	17.92		1.5625	

Example 2: Applying the Properties of Exponents to Rewrite Expressions

Rewrite each expression in the form kx^n, where k is a real number, n is an integer, and x is a nonzero real number.

a. $(5x^5) \cdot (-3x^2)$

b. $\dfrac{3x^5}{(2x)^4}$

c. $\dfrac{3}{(x^2)^{-3}}$

d. $\dfrac{x^{-3}x^4}{x^8}$

Exercises 1–5

Rewrite each expression in the form kx^n, where k is a real number and n is an integer. Assume $x \neq 0$.

1. $2x^5 \cdot x^{10}$

2. $\dfrac{1}{3x^8}$

3. $\dfrac{6x^{-5}}{x^{-3}}$

4. $\left(\dfrac{3}{x^{-22}}\right)^{-3}$

5. $\left(x^2\right)^n \cdot x^3$

EUREKA
MATH™

©2015 Great Minds. eureka-math.org
ALG II-M3-SE-B2-1.3.0-08.2015

Lesson Summary

The Properties of Exponents

For real numbers x and y with $x \neq 0$, $y \neq 0$, and all integers a and b, the following properties hold.

1. $x^a \cdot x^b = x^{a+b}$
2. $(x^a)^b = x^{ab}$
3. $(xy)^a = x^a y^a$
4. $\frac{1}{x^a} = x^{-a}$
5. $\frac{x^a}{x^b} = x^{a-b}$
6. $\left(\frac{x}{y}\right)^a = \frac{x^a}{y^a}$
7. $x^0 = 1$

Problem Set

1. Suppose your class tried to fold an unrolled roll of toilet paper. It was originally 4 in. wide and 30 ft. long. Toilet paper is approximately 0.002 in. thick.

   a. Complete each table, and represent the area and thickness using powers of 2.

Number of Folds n	Thickness After n Folds (in.)
0	
1	
2	
3	
4	
5	
6	

Number of Folds n	Area on Top After n Folds (in^2)
0	
1	
2	
3	
4	
5	
6	

   b. Create an algebraic function that describes the area in square inches after n folds.

   c. Create an algebraic function that describes the thickness in inches after n folds.

2. In the Exit Ticket, we saw the formulas below. The first formula determines the minimum width, W, of a square piece of paper of thickness T needed to fold it in half n times, alternating horizontal and vertical folds. The second formula determines the minimum length, L, of a long rectangular piece of paper of thickness T needed to fold it in half n times, always folding perpendicular to the long side.

$$W = \pi \cdot T \cdot 2^{\frac{3(n-1)}{2}} \qquad L = \frac{\pi T}{6}(2^n + 4)(2^n - 1)$$

Use the appropriate formula to verify why it is possible to fold a 10 inch by 10 inch sheet of gold foil in half 13 times. Use 0.28 millionth of a meter for the thickness of gold foil.

3. Use the formula from Problem 2 to determine if you can fold an unrolled roll of toilet paper in half more than 10 times. Assume that the thickness of a sheet of toilet paper is approximately 0.002 in. and that one roll is 102 ft. long.

4. Apply the properties of exponents to rewrite each expression in the form kx^n, where n is an integer and $x \neq 0$.

   a. $(2x^3)(3x^5)(6x)^2$

   b. $\dfrac{3x^4}{(-6x)^{-2}}$

   c. $\dfrac{x^{-3}x^5}{3x^4}$

   d. $5(x^3)^{-3}(2x)^{-4}$

   e. $\left(\dfrac{x^2}{4x^{-1}}\right)^{-3}$

5. Apply the properties of exponents to verify that each statement is an identity.

   a. $\dfrac{2^{n+1}}{3^n} = 2\left(\dfrac{2}{3}\right)^n$ for integer values of n

   b. $3^{n+1} - 3^n = 2 \cdot 3^n$ for integer values of n

   c. $\dfrac{1}{(3^n)^2} \cdot \dfrac{4^n}{3} = \dfrac{1}{3}\left(\dfrac{2}{3}\right)^{2n}$ for integer values of n

6. Jonah was trying to rewrite expressions using the properties of exponents and properties of algebra for nonzero values of x. In each problem, he made a mistake. Explain where he made a mistake in each part, and provide a correct solution.

Jonah's Incorrect Work
a. $(3x^2)^{-3} = -9x^{-6}$
b. $\dfrac{2}{3x^{-5}} = 6x^5$
c. $\dfrac{2x - x^3}{3x} = \dfrac{2}{3} - x^3$

EUREKA MATH™

7. If $x = 5a^4$ and $a = 2b^3$, express x in terms of b.

8. If $a = 2b^3$ and $b = -\frac{1}{2}c^{-2}$, express a in terms of c.

9. If $x = 3y^4$ and $y = \frac{s}{2x^3}$, show that $s = 54y^{13}$.

10. Do the following tasks without a calculator.

    a. Express 8^3 as a power of 2.

    b. Divide 4^{15} by 2^{10}.

11. Use powers of 2 to perform each calculation without a calculator or other technology.

    a. $\dfrac{2^7 \cdot 2^5}{16}$

    b. $\dfrac{512000}{320}$

12. Write the first five terms of each of the following recursively defined sequences:

    a. $a_{n+1} = 2a_n$, $a_1 = 3$

    b. $a_{n+1} = (a_n)^2$, $a_1 = 3$

    c. $a_{n+1} = 2(a_n)^2$, $a_1 = x$, where x is a real number   Write each term in the form kx^n.

    d. $a_{n+1} = 2(a_n)^{-1}$, $a_1 = y$, $(y \neq 0)$ Write each term in the form kx^n.

13. In Module 1, you established the identity $(1 - r)(1 + r + r^2 + \cdots + r^{n-1}) = 1 - r^n$, where r is a real number and n is a positive integer.

    Use this identity to respond to parts (a)–(g) below.

    a. Rewrite the given identity to isolate the sum $1 + r + r^2 + \cdots + r^{n-1}$ for $r \neq 1$.

    b. Find an explicit formula for $1 + 2 + 2^2 + 2^3 + \cdots + 2^{10}$.

    c. Find an explicit formula for $1 + a + a^2 + a^3 + \cdots + a^{10}$ in terms of powers of a.

    d. Jerry simplified the sum $1 + a + a^2 + a^3 + a^4 + a^5$ by writing $1 + a^{15}$. What did he do wrong?

    e. Find an explicit formula for $1 + 2a + (2a)^2 + (2a)^3 + \cdots + (2a)^{12}$ in terms of powers of a.

    f. Find an explicit formula for $3 + 3(2a) + 3(2a)^2 + 3(2a)^3 + \cdots + 3(2a)^{12}$ in terms of powers of a. Hint: Use part (e).

    g. Find an explicit formula for $P + P(1 + r) + P(1 + r)^2 + P(1 + r)^3 + \cdots + P(1 + r)^{n-1}$ in terms of powers of $(1 + r)$.

This page intentionally left blank

Lesson 2: Base 10 and Scientific Notation

Classwork

Opening Exercise

In the last lesson, you worked with the thickness of a sheet of gold foil (a very small number) and some very large numbers that gave the size of a piece of paper that actually could be folded in half more than 13 times.

a. Convert 0.28 millionth of a meter to centimeters, and express your answer as a decimal number.

b. The length of a piece of square notebook paper that can be folded in half 13 times is 3,294.2 in. Use this number to calculate the area of a square piece of paper that can be folded in half 14 times. Round your answer to the nearest million.

c. Match the equivalent expressions without using a calculator.

$$3.5 \times 10^5 \qquad -6 \qquad -6 \times 10^0 \qquad 0.6 \qquad 3.5 \times 10^{-6}$$
$$3,500,000 \qquad 350,000 \qquad 6 \times 10^{-1} \qquad 0.0000035 \qquad 3.5 \times 10^6$$

©2015 Great Minds. eureka-math.org
ALG II-M3-SE-B2-1.3.0-08.2015

Example 1

Write each number as a product of a decimal number between 1 and 10 and a power of 10.

a. 234,000

b. 0.0035

c. 532,100,000

d. 0.0000000012

e. 3.331

A positive, finite decimal s is said to be written in scientific notation if it is expressed as a product $d \times 10^n$, where d is a finite decimal number so that $1 \leq d < 10$, and n is an integer.

The integer n is called the *order of magnitude* of the decimal $d \times 10^n$.

Exercises 1–6

For Exercises 1–6, write each number in scientific notation.

1. 532,000,000

2. 0.0000000000000000123 (16 zeros after the decimal place)

3. 8,900,000,000,000,000 (14 zeros after the 9)

---

4. 0.00003382

5. 34,000,000,000,000,000,000,000,000 (24 zeros after the 4)

6. 0.000000000000000000000004 (21 zeros after the decimal place)

Exercises 7–8

7. Use the fact that the average distance between the sun and Earth is 151,268,468 km, the average distance between the sun and Jupiter is 780,179,470 km, and the average distance between the sun and Pluto is 5,908,039,124 km to approximate the following distances. Express your answers in scientific notation ($d \times 10^n$), where d is rounded to the nearest tenth.

   a. Distance from the sun to Earth:

   b. Distance from the sun to Jupiter:

   c. Distance from the sun to Pluto:

   d. How much farther Jupiter is from the sun than Earth is from the sun:

   e. How much farther Pluto is from the sun than Jupiter is from the sun:

8. Order the numbers in Exercise 7 from smallest to largest. Explain how writing the numbers in scientific notation helps you to quickly compare and order them.

Example 2: Arithmetic Operations with Numbers Written Using Scientific Notation

a. $(2.4 \times 10^{20}) + (4.5 \times 10^{21})$

b. $(7 \times 10^{-9})(5 \times 10^{5})$

c. $\dfrac{1.2 \times 10^{15}}{3 \times 10^{7}}$

Exercises 9–10

9. Perform the following calculations without rewriting the numbers in decimal form.

   a. $(1.42 \times 10^{15}) - (2 \times 10^{13})$

   b. $(1.42 \times 10^{15})(2.4 \times 10^{13})$

   c. $\dfrac{1.42 \times 10^{-5}}{2 \times 10^{13}}$

10. Estimate how many times farther Jupiter is from the sun than Earth is from the sun. Estimate how many times farther Pluto is from the sun than Earth is from the sun.

Problem Set

1. Write the following numbers used in these statements in scientific notation. (Note: Some of these numbers have been rounded.)

   a. The density of helium is 0.0001785 gram per cubic centimeter.

   b. The boiling point of gold is 5,200°F.

   c. The speed of light is 186,000 miles per second.

   d. One second is 0.000278 hour.

   e. The acceleration due to gravity on the sun is 900 ft/s^2.

   f. One cubic inch is 0.0000214 cubic yard.

   g. Earth's population in 2012 was 7,046,000,000 people.

   h. Earth's distance from the sun is 93,000,000 miles.

   i. Earth's radius is 4,000 miles.

   j. The diameter of a water molecule is 0.000000028 cm.

2. Write the following numbers in decimal form. (Note: Some of these numbers have been rounded.)

   a. A light year is 9.46×10^{15} m.

   b. Avogadro's number is 6.02×10^{23} mol^{-1}.

   c. The universal gravitational constant is 6.674×10^{-11} N $\left(\dfrac{m}{kg}\right)^2$.

   d. Earth's age is 4.54×10^9 years.

   e. Earth's mass is 5.97×10^{24} kg.

   f. A foot is 1.9×10^{-4} mile.

   g. The population of China in 2014 was 1.354×10^9 people.

   h. The density of oxygen is 1.429×10^{-4} gram per liter.

   i. The width of a pixel on a smartphone is 7.8×10^{-2} mm.

   j. The wavelength of light used in optic fibers is 1.55×10^{-6} m.

3. State the necessary value of n that will make each statement true.

   a. $0.000\ 027 = 2.7 \times 10^n$

   b. $-3.125 = -3.125 \times 10^n$

   c. $7{,}540{,}000{,}000 = 7.54 \times 10^n$

   d. $0.033 = 3.3 \times 10^n$

   e. $15 = 1.5 \times 10^n$

   f. $26{,}000 \times 200 = 5.2 \times 10^n$

   g. $3000 \times 0.0003 = 9 \times 10^n$

   h. $0.0004 \times 0.002 = 8 \times 10^n$

   i. $\dfrac{16000}{80} = 2 \times 10^n$

EUREKA
MATH™

j. $\dfrac{500}{0.002} = 2.5 \times 10^n$

4. Perform the following calculations without rewriting the numbers in decimal form.

   a. $(2.5 \times 10^4) + (3.7 \times 10^3)$

   b. $(6.9 \times 10^{-3}) - (8.1 \times 10^{-3})$

   c. $(6 \times 10^{11})(2.5 \times 10^{-5})$

   d. $\dfrac{4.5 \times 10^8}{2 \times 10^{10}}$

5. The wavelength of visible light ranges from 650 nanometers to 850 nanometers, where $1 \text{ nm} = 1 \times 10^{-7}$ cm. Express the range of wavelengths of visible light in centimeters.

6. In 1694, the Dutch scientist Antonie van Leeuwenhoek was one of the first scientists to see a red blood cell in a microscope. He approximated that a red blood cell was "25,000 times as small as a grain of sand." Assume a grain of sand is $\dfrac{1}{2}$ mm wide, and a red blood cell is approximately 7 micrometers wide. One micrometer is 1×10^{-6} m. Support or refute Leeuwenhoek's claim. Use scientific notation in your calculations.

7. When the Mars Curiosity Rover entered the atmosphere of Mars on its descent in 2012, it was traveling roughly 13,200 mph. On the surface of Mars, its speed averaged 0.00073 mph. How many times faster was the speed when it entered the atmosphere than its typical speed on the planet's surface? Use scientific notation in your calculations.

8. Earth's surface is approximately 70% water. There is no water on the surface of Mars, and its diameter is roughly half of Earth's diameter. Assume both planets are spherical. The radius of Earth is approximately 4,000 miles. The surface area of a sphere is given by the formula $SA = 4\pi r^2$, where r is the radius of the sphere. Which has more land mass, Earth or Mars? Use scientific notation in your calculations.

9. There are approximately 25 trillion (2.5×10^{13}) red blood cells in the human body at any one time. A red blood cell is approximately 7×10^{-6} m wide. Imagine if you could line up all your red blood cells end to end. How long would the line of cells be? Use scientific notation in your calculations.

10. Assume each person needs approximately 100 square feet of living space. Now imagine that we are going to build a giant apartment building that will be 1 mile wide and 1 mile long to house all the people in the United States, estimated to be 313.9 million people in 2012. If each floor of the apartment building is 10 feet high, how tall will the apartment building be?

This page intentionally left blank

Lesson 3: Rational Exponents—What are $2^{\frac{1}{2}}$ and $2^{\frac{1}{3}}$?

Classwork

Opening Exercise

a. What is the value of $2^{\frac{1}{2}}$? Justify your answer.

b. Graph $f(x) = 2^x$ for each integer x from $x = -2$ to $x = 5$. Connect the points on your graph with a smooth curve.

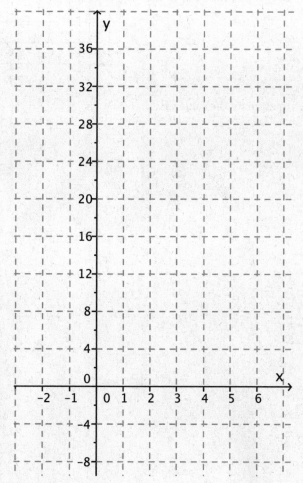

Lesson 3:   Rational Exponents—What Are $2^{\frac{1}{2}}$ and $2^{\frac{1}{3}}$?

S.17

©2015 Great Minds. eureka-math.org
ALG II-M3-SE-B2-1.3.0-08.2015

The graph on the right shows a close-up view of $f(x) = 2^x$ for $-0.5 < x < 1.5$.

c.   Find two consecutive integers that are under and over estimates of the value of $2^{\frac{1}{2}}$.

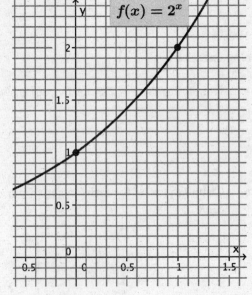

d.   Does it appear that $2^{\frac{1}{2}}$ is halfway between the integers you specified in part (c)?

e.   Use the graph of $f(x) = 2^x$ to estimate the value of $2^{\frac{1}{2}}$.

f.   Use the graph of $f(x) = 2^x$ to estimate the value of $2^{\frac{1}{3}}$.

Example

a.   What is the 4ᵗʰ root of 16?

b.   What is the cube root of 125?

c.   What is the 5ᵗʰ root of 100,000?

Lesson 3:   Rational Exponents—What Are $2^{\frac{1}{2}}$ and $2^{\frac{1}{3}}$?

EUREKA MATH™

Exercise 1

Evaluate each expression.

a. $\sqrt[4]{81}$

b. $\sqrt[5]{32}$

c. $\sqrt[3]{9} \cdot \sqrt[3]{3}$

d. $\sqrt[4]{25} \cdot \sqrt[4]{100} \cdot \sqrt[4]{4}$

Discussion

If $2^{\frac{1}{2}} = \sqrt{2}$ and $2^{\frac{1}{3}} = \sqrt[3]{2}$, what does $2^{\frac{3}{4}}$ equal? Explain your reasoning.

Exercises 2–12

Rewrite each exponential expression as a radical expression.

2. $3^{\frac{1}{2}}$

3. $11^{\frac{1}{5}}$

EUREKA MATH™

Lesson 3: Rational Exponents—What Are $2^{\frac{1}{2}}$ and $2^{\frac{1}{3}}$?

S.19

4. $\left(\dfrac{1}{4}\right)^{\frac{1}{5}}$

5. $6^{\frac{1}{10}}$

Rewrite the following exponential expressions as equivalent radical expressions. If the number is rational, write it without radicals or exponents.

6. $2^{\frac{3}{2}}$

7. $4^{\frac{5}{2}}$

8. $\left(\dfrac{1}{8}\right)^{\frac{5}{3}}$

9. Show why the following statement is true:

$$2^{-\frac{1}{2}} = \dfrac{1}{2^{\frac{1}{2}}}$$

Lesson 3: Rational Exponents—What Are $2^{\frac{1}{2}}$ and $2^{\frac{1}{3}}$?

EUREKA MATH

Rewrite the following exponential expressions as equivalent radical expressions. If the number is rational, write it without radicals or exponents.

10. $4^{-\frac{3}{2}}$

11. $27^{-\frac{2}{3}}$

12. $\left(\frac{1}{4}\right)^{-\frac{1}{2}}$

EUREKA
MATH™

Lesson 3:   Rational Exponents—What Are $2^{\frac{1}{2}}$ and $2^{\frac{1}{3}}$?

S.21

©2015 Great Minds. eureka-math.org
ALG II-M3-SE-B2-1.3.0-08.2015

Lesson Summary

n^{TH} **ROOT OF A NUMBER:** Let a and b be numbers, and let n be a positive integer. If $b = a^n$, then a is a nth root of b. If $n = 2$, then the root is a called a *square root*. If $n = 3$, then the root is called a *cube root*.

PRINCIPAL n^{TH} ROOT OF A NUMBER: Let b be a real number that has at least one real n^{th} root. The *principal n^{th} root of b* is the real n^{th} root that has the same sign as b and is denoted by a radical symbol: $\sqrt[n]{b}$.

Every positive number has a unique principal n^{th} root. We often refer to the principal n^{th} root of b as just the nth root of b. The nth root of 0 is 0.

For any positive integers m and n, and any real number b for which the principal n^{th} root of b exists, we have

$$b^{\frac{1}{n}} = \sqrt[n]{b}$$

$$b^{\frac{m}{n}} = \sqrt[n]{b^m} = \left(\sqrt[n]{b}\right)^m$$

$$b^{-\frac{m}{n}} = \frac{1}{\sqrt[n]{b^m}} \text{ for } b \neq 0.$$

Problem Set

1. Select the expression from (A), (B), and (C) that correctly completes the statement.

		(A)	(B)	(C)
a.	$x^{\frac{1}{3}}$ is equivalent to _____.	$\frac{1}{3}x$	$\sqrt[3]{x}$	$\frac{3}{x}$
b.	$x^{\frac{2}{3}}$ is equivalent to _____.	$\frac{2}{3}x$	$\sqrt[3]{x^2}$	$\left(\sqrt{x}\right)^3$
c.	$x^{-\frac{1}{4}}$ is equivalent to _____.	$-\frac{1}{4}x$	$\frac{4}{x}$	$\frac{1}{\sqrt[4]{x}}$
d.	$\left(\frac{4}{x}\right)^{\frac{1}{2}}$ is equivalent to _____.	$\frac{2}{x}$	$\frac{4}{x^2}$	$\frac{2}{\sqrt{x}}$

2. Identify which of the expressions (A), (B), and (C) are equivalent to the given expression.

		(A)	(B)	(C)
a.	$16^{\frac{1}{2}}$	$\left(\frac{1}{16}\right)^{-\frac{1}{2}}$	$8^{\frac{2}{3}}$	$64^{\frac{3}{2}}$
b.	$\left(\frac{2}{3}\right)^{-1}$	$-\frac{3}{2}$	$\left(\frac{9}{4}\right)^{\frac{1}{2}}$	$\frac{27^{\frac{1}{3}}}{6}$

Lesson 3:   Rational Exponents—What Are $2^{\frac{1}{2}}$ and $2^{\frac{1}{3}}$?

EUREKA
MATH™

3. Rewrite in radical form. If the number is rational, write it without using radicals.

    a. $6^{\frac{3}{2}}$

    b. $\left(\frac{1}{2}\right)^{\frac{1}{4}}$

    c. $3(8)^{\frac{1}{3}}$

    d. $\left(\frac{64}{125}\right)^{-\frac{2}{3}}$

    e. $81^{-\frac{1}{4}}$

4. Rewrite the following expressions in exponent form.

    a. $\sqrt{5}$

    b. $\sqrt[3]{5^2}$

    c. $\sqrt{5^3}$

    d. $\left(\sqrt[3]{5}\right)^2$

5. Use the graph of $f(x) = 2^x$ shown to the right to estimate the following powers of 2.

    a. $2^{\frac{1}{4}}$

    b. $2^{\frac{2}{3}}$

    c. $2^{\frac{3}{4}}$

    d. $2^{0.2}$

    e. $2^{1.2}$

    f. $2^{-\frac{1}{5}}$

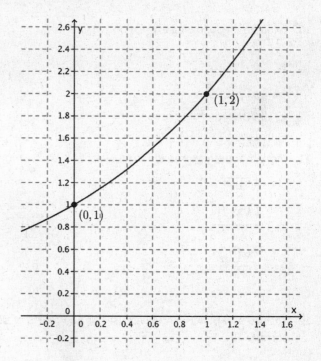

EUREKA MATH™

Lesson 3:   Rational Exponents—What Are $2^{\frac{1}{2}}$ and $2^{\frac{1}{3}}$?

S.23

©2015 Great Minds. eureka-math.org
ALG II-M3-SE-B2-1.3.0-08.2015

6. Rewrite each expression in the form kx^n, where k is a real number, x is a positive real number, and n is rational.

   a. $\sqrt[4]{16x^3}$

   b. $\dfrac{5}{\sqrt{x}}$

   c. $\sqrt[3]{1/x^4}$

   d. $\dfrac{4}{\sqrt[3]{8x^3}}$

   e. $\dfrac{27}{\sqrt{9x^4}}$

   f. $\left(\dfrac{125}{x^2}\right)^{-\frac{1}{3}}$

7. Find the value of x for which $2x^{\frac{1}{2}} = 32$.

8. Find the value of x for which $x^{\frac{4}{3}} = 81$.

9. If $x^{\frac{3}{2}} = 64$, find the value of $4x^{-\frac{3}{4}}$.

10. Evaluate the following expressions when $b = \dfrac{1}{9}$.

    a. $b^{-\frac{1}{2}}$

    b. $b^{\frac{5}{2}}$

    c. $\sqrt[3]{3b^{-1}}$

11. Show that each expression is equivalent to $2x$. Assume x is a positive real number.

    a. $\sqrt[4]{16x^4}$

    b. $\dfrac{\left(\sqrt[3]{8x^3}\right)^2}{\sqrt{4x^2}}$

    c. $\dfrac{6x^3}{\sqrt[3]{27x^6}}$

12. Yoshiko said that $16^{\frac{1}{4}} = 4$ because 4 is one-fourth of 16. Use properties of exponents to explain why she is or is not correct.

13. Jefferson said that $8^{\frac{4}{3}} = 16$ because $8^{\frac{1}{3}} = 2$ and $2^4 = 16$. Use properties of exponents to explain why he is or is not correct.

S.24

Lesson 3: Rational Exponents—What Are $2^{\frac{1}{2}}$ and $2^{\frac{1}{3}}$?

EUREKA MATH™

14. Rita said that $8^{\frac{2}{3}} = 128$ because $8^{\frac{2}{3}} = 8^2 \cdot 8^{\frac{1}{3}}$, so $8^{\frac{2}{3}} = 64 \cdot 2$, and then $8^{\frac{2}{3}} = 128$. Use properties of exponents to explain why she is or is not correct.

15. Suppose for some positive real number a that $\left(a^{\frac{1}{4}} \cdot a^{\frac{1}{2}} \cdot a^{\frac{1}{4}}\right)^2 = 3$.

    a.   What is the value of a?

    b.   Which exponent properties did you use to find your answer to part (a)?

16. In the lesson, you made the following argument:

$$\left(2^{\frac{1}{3}}\right)^3 = 2^{\frac{1}{3}} \cdot 2^{\frac{1}{3}} \cdot 2^{\frac{1}{3}}$$
$$= 2^{\frac{1}{3}+\frac{1}{3}+\frac{1}{3}}$$
$$= 2^1$$
$$= 2.$$

Since $\sqrt[3]{2}$ is a number so that $\left(\sqrt[3]{2}\right)^3 = 2$ and $2^{\frac{1}{3}}$ is a number so that $\left(2^{\frac{1}{3}}\right)^3 = 2$, you concluded that $2^{\frac{1}{3}} = \sqrt[3]{2}$.

Which exponent property was used to make this argument?

EUREKA
MATH™

Lesson 3:   Rational Exponents—What Are $2^{\frac{1}{2}}$ and $2^{\frac{1}{3}}$?

S.25

This page intentionally left blank

Lesson 4: Properties of Exponents and Radicals

Classwork

Opening Exercise

Write each exponential expression as a radical expression, and then use the definition and properties of radicals to write the resulting expression as an integer.

a. $7^{\frac{1}{2}} \cdot 7^{\frac{1}{2}}$

b. $3^{\frac{1}{3}} \cdot 3^{\frac{1}{3}} \cdot 3^{\frac{1}{3}}$

c. $12^{\frac{1}{2}} \cdot 3^{\frac{1}{2}}$

d. $\left(64^{\frac{1}{3}}\right)^{\frac{1}{2}}$

Examples 1–3

Write each expression in the form $b^{\frac{m}{n}}$ for positive real numbers b and integers m and n with $n > 0$ by applying the properties of radicals and the definition of nth root.

1. $b^{\frac{1}{4}} \cdot b^{\frac{1}{4}}$

2. $b^{\frac{1}{3}} \cdot b^{\frac{4}{3}}$

EUREKA
MATH™

3. $b^{\frac{1}{5}} \cdot b^{\frac{3}{4}}$

Exercises 1–4

Write each expression in the form $b^{\frac{m}{n}}$. If a numeric expression is a rational number, then write your answer without exponents.

1. $b^{\frac{2}{3}} \cdot b^{\frac{1}{2}}$

2. $\left(b^{-\frac{1}{5}}\right)^{\frac{2}{3}}$

3. $64^{\frac{1}{3}} \cdot 64^{\frac{3}{2}}$

4. $\left(\dfrac{9^3}{4^2}\right)^{\frac{3}{2}}$

Example 4

Rewrite the radical expression $\sqrt{48x^5y^4z^2}$ so that no perfect square factors remain inside the radical.

Exercise 5

5. Use the definition of rational exponents and properties of exponents to rewrite each expression with rational exponents containing as few fractions as possible. Then, evaluate each resulting expression for $x = 50$, $y = 12$, and $z = 3$.

    a. $\sqrt{8x^3y^2}$

EUREKA
MATH™

b. $\sqrt[3]{54y^7z^2}$

Exercise 6

6. Order these numbers from smallest to largest. Explain your reasoning.

$16^{2.5}$ $\qquad\qquad\qquad$ $9^{3.6}$ $\qquad\qquad\qquad$ $32^{1.2}$

Lesson Summary

The properties of exponents developed in Grade 8 for integer exponents extend to rational exponents.

That is, for any integers m, n, p, and q, with $n > 0$ and $q > 0$, and any real numbers a and b so that $a^{\frac{1}{n}}$, $b^{\frac{1}{n}}$, and $b^{\frac{1}{q}}$ are defined, we have the following properties of exponents:

1. $b^{\frac{m}{n}} \cdot b^{\frac{p}{q}} = b^{\frac{m}{n}+\frac{p}{q}}$

2. $b^{\frac{m}{n}} = \sqrt[n]{b^m}$

3. $\left(b^{\frac{1}{n}}\right)^n = b$

4. $\left(b^n\right)^{\frac{1}{n}} = b$

5. $(ab)^{\frac{m}{n}} = a^{\frac{m}{n}} \cdot b^{\frac{m}{n}}$

6. $\left(b^{\frac{m}{n}}\right)^{\frac{p}{q}} = b^{\frac{mp}{nq}}$

7. $b^{-\frac{m}{n}} = \frac{1}{b^{\frac{m}{n}}}$.

Problem Set

1. Evaluate each expression for $a = 27$ and $b = 64$.

    a. $\sqrt[3]{a}\sqrt{b}$

    b. $\left(3\sqrt[3]{a}\sqrt{b}\right)^2$

    c. $\left(\sqrt[3]{a} + 2\sqrt{b}\right)^2$

    d. $a^{-\frac{2}{3}} + b^{\frac{3}{2}}$

    e. $\left(a^{-\frac{2}{3}} \cdot b^{\frac{3}{2}}\right)^{-1}$

    f. $\left(a^{-\frac{2}{3}} - \frac{1}{8}b^{\frac{3}{2}}\right)^{-1}$

2. Rewrite each expression so that each term is in the form kx^n, where k is a real number, x is a positive real number, and n is a rational number.

    a. $x^{-\frac{2}{3}} \cdot x^{\frac{1}{3}}$

    b. $2x^{\frac{1}{2}} \cdot 4x^{-\frac{5}{2}}$

    c. $\dfrac{10x^{\frac{1}{3}}}{2x^2}$

    d. $\left(3x^{\frac{1}{4}}\right)^{-2}$

    e. $x^{\frac{1}{2}}\left(2x^2 - \frac{4}{x}\right)$

    f. $\sqrt[3]{\dfrac{27}{x^6}}$

    g. $\sqrt[3]{x} \cdot \sqrt[3]{-8x^2} \cdot \sqrt[3]{27x^4}$

    h. $\dfrac{2x^4 - x^2 - 3x}{\sqrt{x}}$

    i. $\dfrac{\sqrt{x} - 2x^{-3}}{4x^2}$

Lesson 4:   Properties of Exponents and Radicals

EUREKA MATH™

©2015 Great Minds. eureka-math.org
ALG II-M3-SE-B2-1.3.0-08.2015

3. Show that $\left(\sqrt{x} + \sqrt{y}\right)^2$ is not equal to $x^1 + y^1$ when $x = 9$ and $y = 16$.

4. Show that $\left(x^{\frac{1}{2}} + y^{\frac{1}{2}}\right)^{-1}$ is not equal to $\dfrac{1}{x^{\frac{1}{2}}} + \dfrac{1}{y^{\frac{1}{2}}}$ when $x = 9$ and $y = 16$.

5. From these numbers, select (a) one that is negative, (b) one that is irrational, (c) one that is not a real number, and (d) one that is a perfect square:

$$3^{\frac{1}{2}} \cdot 9^{\frac{1}{2}}, \; 27^{\frac{1}{3}} \cdot 144^{\frac{1}{2}}, \; 64^{\frac{1}{3}} - 64^{\frac{2}{3}}, \text{ and } \left(4^{-\frac{1}{2}} - 4^{\frac{1}{2}}\right)^{\frac{1}{2}}$$

6. Show that for any rational number n, the expression $2^n \cdot 4^{n+1} \cdot \left(\frac{1}{8}\right)^n$ is equal to 4.

7. Let n be any rational number. Express each answer as a power of 10.
   a. Multiply 10^n by 10.
   b. Multiply $\sqrt{10}$ by 10^n.
   c. Square 10^n.
   d. Divide $100 \cdot 10^n$ by 10^{2n}.
   e. Show that $10^n = 11 \cdot 10^n - 10^{n+1}$.

8. Rewrite each of the following radical expressions as an equivalent exponential expression in which each variable occurs no more than once.
   a. $\sqrt{8x^2y}$
   b. $\sqrt[5]{96x^3y^{15}z^6}$

9. Use properties of exponents to find two integers that are upper and lower estimates of the value of $4^{1.6}$.

10. Use properties of exponents to find two integers that are upper and lower estimates of the value of $8^{2.3}$.

11. Kepler's third law of planetary motion relates the average distance, a, of a planet from the sun to the time, t, it takes the planet to complete one full orbit around the sun according to the equation $t^2 = a^3$. When the time, t, is measured in Earth years, the distance, a, is measured in astronomical units (AUs). (One AU is equal to the average distance from Earth to the sun.)
   a. Find an equation for t in terms of a and an equation for a in terms of t.
   b. Venus takes about 0.616 Earth year to orbit the sun. What is its average distance from the sun?
   c. Mercury is an average distance of 0.387 AU from the sun. About how long is its orbit in Earth years?

This page intentionally left blank

Lesson 5: Irrational Exponents—What are $2^{\sqrt{2}}$ and 2^{π}?

Classwork

Exercise 1

a. Write the following finite decimals as fractions (you do not need to reduce to lowest terms).

$$1, \quad 1.4, \quad 1.41, \quad 1.414, \quad 1.4142, \quad 1.41421$$

b. Write $2^{1.4}$, $2^{1.41}$, $2^{1.414}$, and $2^{1.4142}$ in radical form ($\sqrt[n]{2^m}$).

c. Use a calculator to compute decimal approximations of the radical expressions you found in part (b) to 5 decimal places. For each approximation, underline the digits that are also in the previous approximation, starting with 2.00000 done for you below. What do you notice?

$$2^1 = 2 = 2.00000$$

Lesson 5:   Irrational Exponents—What are $2^{\sqrt{2}}$ and 2^{π}?

S.35

Exercise 2

a. Write six terms of a sequence that a calculator can use to approximate 2^π.
   (Hint: $\pi = 3.141\,59\,\ldots$)

b. Compute $2^{3.14}$ and 2^π on your calculator. In which digit do they start to differ?

c. How could you improve the accuracy of your estimate of 2^π?

EUREKA
MATH

Problem Set

1. Is it possible for a number to be both rational and irrational?

2. Use properties of exponents to rewrite the following expressions as a number or an exponential expression with only one exponent.

   a. $\left(2^{\sqrt{3}}\right)^{\sqrt{3}}$

   b. $\left(\sqrt{2}^{\sqrt{2}}\right)^{\sqrt{2}}$

   c. $\left(3^{1+\sqrt{5}}\right)^{1-\sqrt{5}}$

   d. $3^{\frac{1+\sqrt{5}}{2}} \cdot 3^{\frac{1-\sqrt{5}}{2}}$

   e. $3^{\frac{1+\sqrt{5}}{2}} \div 3^{\frac{1-\sqrt{5}}{2}}$

   f. $3^{2\cos^2(x)} \cdot 3^{2\sin^2(x)}$

3.

   a. Between what two integer powers of 2 does $2^{\sqrt{5}}$ lie?

   b. Between what two integer powers of 3 does $3^{\sqrt{10}}$ lie?

   c. Between what two integer powers of 5 does $5^{\sqrt{3}}$ lie?

4. Use the process outlined in the lesson to approximate the number $2^{\sqrt{5}}$. Use the approximation $\sqrt{5} \approx 2.236\,067\,98$.

   a. Find a sequence of five intervals that contain $\sqrt{5}$ whose endpoints get successively closer to $\sqrt{5}$.

   b. Find a sequence of five intervals that contain $2^{\sqrt{5}}$ whose endpoints get successively closer to $2^{\sqrt{5}}$. Write your intervals in the form $2^r < 2^{\sqrt{5}} < 2^s$ for rational numbers r and s.

   c. Use your calculator to find approximations to four decimal places of the endpoints of the intervals in part (b).

   d. Based on your work in part (c), what is your best estimate of the value of $2^{\sqrt{5}}$?

   e. Can we tell if $2^{\sqrt{5}}$ is rational or irrational? Why or why not?

5. Use the process outlined in the lesson to approximate the number $3^{\sqrt{10}}$. Use the approximation $\sqrt{10} \approx 3.162\,277\,7$.

   a. Find a sequence of five intervals that contain $3^{\sqrt{10}}$ whose endpoints get successively closer to $3^{\sqrt{10}}$. Write your intervals in the form $3^r < 3^{\sqrt{10}} < 3^s$ for rational numbers r and s.

   b. Use your calculator to find approximations to four decimal places of the endpoints of the intervals in part (a).

   c. Based on your work in part (b), what is your best estimate of the value of $3^{\sqrt{10}}$?

EUREKA MATH™

Lesson 5: Irrational Exponents—What are $2^{\sqrt{2}}$ and 2^π?

S.37

6. Use the process outlined in the lesson to approximate the number $5^{\sqrt{7}}$. Use the approximation $\sqrt{7} \approx 2.645\,751\,31$.

   a. Find a sequence of seven intervals that contain $5^{\sqrt{7}}$ whose endpoints get successively closer to $5^{\sqrt{7}}$. Write your intervals in the form $5^r < 5^{\sqrt{7}} < 5^s$ for rational numbers r and s.

   b. Use your calculator to find approximations to four decimal places of the endpoints of the intervals in part (a).

   c. Based on your work in part (b), what is your best estimate of the value of $5^{\sqrt{7}}$?

7. A rational number raised to a rational power can either be rational or irrational. For example, $4^{\frac{1}{2}}$ is rational because $4^{\frac{1}{2}} = 2$, and $2^{\frac{1}{4}}$ is irrational because $2^{\frac{1}{4}} = \sqrt[4]{2}$. In this problem, you will investigate the possibilities for an irrational number raised to an irrational power.

   a. Evaluate $\left(\sqrt{2}\right)^{\left(\sqrt{2}\right)^{\sqrt{2}}}$.

   b. Can the value of an irrational number raised to an irrational power ever be rational?

EUREKA MATH™

Lesson 6: Euler's Number, e

Classwork

Exercises 1–3

1. Assume that there is initially 1 cm of water in the tank, and the height of the water doubles every 10 seconds. Write an equation that could be used to calculate the height $H(t)$ of the water in the tank at any time t.

2. How would the equation in Exercise 1 change if...

   a. the initial depth of water in the tank was 2 cm?

   b. the initial depth of water in the tank was $\frac{1}{2}$ cm?

   c. the initial depth of water in the tank was 10 cm?

   d. the initial depth of water in the tank was A cm, for some positive real number A?

3. How would the equation in Exercise 2, part (d), change if...

   a. the height tripled every ten seconds?

   b. the height doubled every five seconds?

c. the height quadrupled every second?

d. the height halved every ten seconds?

Example

Consider two identical water tanks, each of which begins with a height of water 1 cm and fills with water at a different rate. Which equations can be used to calculate the height of water in each tank at time t? Use H_1 for tank 1 and H_2 for tank 2.

The height of the water in tank 1 doubles every second.

The height of the water in tank 2 triples every second.

a. If both tanks start filling at the same time, which one fills first?

b. We want to know the average rate of change of the height of the water in these tanks over an interval that starts at a fixed time T as they are filling up. What is the formula for the average rate of change of a function f on an interval $[a, b]$?

c. What is the formula for the average rate of change of the function H_1 on an interval $[a, b]$?

d. Let's calculate the average rate of change of the function H_1 on the interval $[T, T + 0.1]$, which is an interval one-tenth of a second long starting at an unknown time T.

Lesson 6:   Euler's Number, e

Exercises 4–8

4. For the second tank, calculate the average change in the height, H_2, from time T seconds to $T + 0.1$ second. Express the answer as a number times the value of the original function at time T. Explain the meaning of these findings.

5. For each tank, calculate the average change in height from time T seconds to $T + 0.001$ second. Express the answer as a number times the value of the original function at time T. Explain the meaning of these findings.

6. In Exercise 5, the average rate of change of the height of the water in tank 1 on the interval $[T, T + 0.001]$ can be described by the expression $c_1 \cdot 2^T$, and the average rate of change of the height of the water in tank 2 on the interval $[T, T + 0.001]$ can be described by the expression $c_2 \cdot 3^T$. What are approximate values of c_1 and c_2?

7. As an experiment, let's look for a value of b so that if the height of the water can be described by $H(t) = b^t$, then the expression for the average rate of change on the interval $[T, T + 0.001]$ is $1 \cdot H(T)$.

   a. Write out the expression for the average rate of change of $H(t) = b^t$ on the interval $[T, T + 0.001]$.

   b. Set your expression in part (a) equal to $1 \cdot H(T)$, and reduce to an expression involving a single b.

   c. Now we want to find the value of b that satisfies the equation you found in part (b), but we do not have a way to explicitly solve this equation. Look back at Exercise 6; which two consecutive integers have b between them?

   d. Use your calculator and a guess-and-check method to find an approximate value of b to 2 decimal places.

8. Verify that for the value of b found in Exercise 7, $\dfrac{H_b(T + 0.001) - H_b(T)}{0.001} \approx H_b(T)$, where $H_b(T) = b^T$.

EUREKA
MATH

Lesson Summary

- Euler's number, e, is an irrational number that is approximately equal to 2.718 281 828 459 0.
- **AVERAGE RATE OF CHANGE:** Given a function f whose domain contains the interval of real numbers $[a, b]$ and whose range is a subset of the real numbers, the *average rate of change on the interval* $[a, b]$ is defined by the number

$$\frac{f(b) - f(a)}{b - a}.$$

Problem Set

1. The product $4 \cdot 3 \cdot 2 \cdot 1$ is called 4 *factorial* and is denoted by $4!$. Then $10! = 10 \cdot 9 \cdot 8 \cdot 7 \cdot 6 \cdot 5 \cdot 4 \cdot 3 \cdot 2 \cdot 1$, and for any positive integer n, $n! = n(n-1)(n-2) \cdot \cdots \cdot 3 \cdot 2 \cdot 1$.

   a. Complete the following table of factorial values:

n	1	2	3	4	5	6	7	8
$n!$								

   b. Evaluate the sum $1 + \frac{1}{1!}$.

   c. Evaluate the sum $1 + \frac{1}{1!} + \frac{1}{2!}$.

   d. Use a calculator to approximate the sum $1 + \frac{1}{1!} + \frac{1}{2!} + \frac{1}{3!}$ to 7 decimal places. Do not round the fractions before evaluating the sum.

   e. Use a calculator to approximate the sum $1 + \frac{1}{1!} + \frac{1}{2!} + \frac{1}{3!} + \frac{1}{4!}$ to 7 decimal places. Do not round the fractions before evaluating the sum.

   f. Use a calculator to approximate sums of the form $1 + \frac{1}{1!} + \frac{1}{2!} + \cdots + \frac{1}{k!}$ to 7 decimal places for $k = 5, 6, 7, 8, 9, 10$. Do not round the fractions before evaluating the sums with a calculator.

   g. Make a conjecture about the sums $1 + \frac{1}{1!} + \frac{1}{2!} + \cdots + \frac{1}{k!}$ for positive integers k as k increases in size.

   h. Would calculating terms of this sequence ever yield an exact value of e? Why or why not?

2. Consider the sequence given by $a_n = \left(1 + \frac{1}{n}\right)^n$, where $n \geq 1$ is an integer.

   a. Use your calculator to approximate the first 5 terms of this sequence to 7 decimal places.

   b. Does it appear that this sequence settles near a particular value?

   c. Use a calculator to approximate the following terms of this sequence to 7 decimal places.

      i. a_{100}

      ii. a_{1000}

      iii. $a_{10,000}$

      iv. $a_{100,000}$

      v. $a_{1,000,000}$

      vi. $a_{10,000,000}$

      vii. $a_{100,000,000}$

   d. Does it appear that this sequence settles near a particular value?

   e. Compare the results of this exercise with the results of Problem 1. What do you observe?

3. If $x = 5a^4$ and $a = 2e^3$, express x in terms of e, and approximate to the nearest whole number.

4. If $a = 2b^3$ and $b = -\frac{1}{2}e^{-2}$, express a in terms of e, and approximate to four decimal places.

5. If $x = 3e^4$ and $= \frac{s}{2x^3}$, show that $s = 54e^{13}$, and approximate s to the nearest whole number.

6. The following graph shows the number of barrels of oil produced by the Glenn Pool well in Oklahoma from 1910 to 1916.

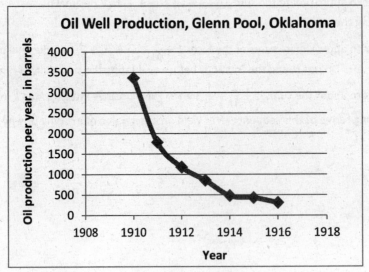

Source: Cutler, Willard W., Jr. Estimation of Underground Oil Reserves by Oil-Well Production Curves, U.S. Department of the Interior, 1924.

Lesson 6:   Euler's Number, e

EUREKA MATH™

a. Estimate the average rate of change of the amount of oil produced by the well on the interval $[1910, 1916]$, and explain what that number represents.

b. Estimate the average rate of change of the amount of oil produced by the well on the interval $[1910, 1913]$, and explain what that number represents.

c. Estimate the average rate of change of the amount of oil produced by the well on the interval $[1913, 1916]$, and explain what that number represents.

d. Compare your results for the rates of change in oil production in the first half and the second half of the time period in question in parts (b) and (c). What do those numbers say about the production of oil from the well?

e. Notice that the average rate of change of the amount of oil produced by the well on any interval starting and ending in two consecutive years is always negative. Explain what that means in the context of oil production.

7. The following table lists the number of hybrid electric vehicles (HEVs) sold in the United States between 1999 and 2013.

Year	Number of HEVs Sold in U.S.	Year	Number of HEVs Sold in U.S.
1999	17	2007	352,274
2000	9350	2008	312,386
2001	20,282	2009	290,271
2002	36,035	2010	274,210
2003	47,600	2011	268,752
2004	84,199	2012	434,498
2005	209,711	2013	495,685
2006	252,636		

Source: U.S. Department of Energy, Alternative Fuels and Advanced Vehicle Data Center, 2013.

a. During which one-year interval is the average rate of change of the number of HEVs sold the largest? Explain how you know.

b. Calculate the average rate of change of the number of HEVs sold on the interval $[2003, 2004]$, and explain what that number represents.

c. Calculate the average rate of change of the number of HEVs sold on the interval $[2003, 2008]$, and explain what that number represents.

d. What does it mean if the average rate of change of the number of HEVs sold is negative?

Extension:

8. The formula for the area of a circle of radius r can be expressed as a function $A(r) = \pi r^2$.

a. Find the average rate of change of the area of a circle on the interval $[4, 5]$.

b. Find the average rate of change of the area of a circle on the interval $[4, 4.1]$.

c. Find the average rate of change of the area of a circle on the interval $[4, 4.01]$.

d. Find the average rate of change of the area of a circle on the interval $[4, 4.001]$.

e. What is happening to the average rate of change of the area of the circle as the interval gets smaller and smaller?

f. Find the average rate of change of the area of a circle on the interval $[4, 4 + h]$ for some small positive number h.

g. What happens to the average rate of change of the area of the circle on the interval $[4, 4 + h]$ as $h \to 0$? Does this agree with your answer to part (d)? Should it agree with your answer to part (e)?

h. Find the average rate of change of the area of a circle on the interval $[r_0, r_0 + h]$ for some positive number r_0 and some small positive number h.

i. What happens to the average rate of change of the area of the circle on the interval $[r_0, r_0 + h]$ as $h \to 0$? Do you recognize the resulting formula?

9. The formula for the volume of a sphere of radius r can be expressed as a function $V(r) = \frac{4}{3}\pi r^3$. As you work through these questions, you will see the pattern develop more clearly if you leave your answers in the form of a coefficient times π. Approximate the coefficient to five decimal places.

a. Find the average rate of change of the volume of a sphere on the interval $[2, 3]$.

b. Find the average rate of change of the volume of a sphere on the interval $[2, 2.1]$.

c. Find the average rate of change of the volume of a sphere on the interval $[2, 2.01]$.

d. Find the average rate of change of the volume of a sphere on the interval $[2, 2.001]$.

e. What is happening to the average rate of change of the volume of a sphere as the interval gets smaller and smaller?

f. Find the average rate of change of the volume of a sphere on the interval $[2, 2 + h]$ for some small positive number h.

g. What happens to the average rate of change of the volume of a sphere on the interval $[2, 2 + h]$ as $h \to 0$? Does this agree with your answer to part (e)? Should it agree with your answer to part (e)?

h. Find the average rate of change of the volume of a sphere on the interval $[r_0, r_0 + h]$ for some positive number r_0 and some small positive number h.

i. What happens to the average rate of change of the volume of a sphere on the interval $[r_0, r_0 + h]$ as $h \to 0$? Do you recognize the resulting formula?

Lesson 7: Bacteria and Exponential Growth

Classwork

Opening Exercise

Work with your partner or group to solve each of the following equations for x.

   a. $2^x = 2^3$

   b. $2^x = 2$

   c. $2^x = 16$

   d. $2^x - 64 = 0$

   e. $2^x - 1 = 0$

   f. $2^{3x} = 64$

   g. $2^{x+1} = 32$

Example

The *Escherichia coli* bacteria (commonly known as *E. coli*) reproduces once every 30 minutes, meaning that a colony of *E. coli* can double every half hour. *Mycobacterium tuberculosis* has a generation time in the range of 12 to 16 hours. Researchers have found evidence that suggests certain bacteria populations living deep below the surface of the earth may grow at extremely slow rates, reproducing once every several thousand years. With this variation in bacterial growth rates, it is reasonable that we assume a 24-hour reproduction time for a hypothetical bacteria colony in this example.

Suppose we have a bacteria colony that starts with 1 bacterium, and the population of bacteria doubles every day.

What function P can we use to model the bacteria population on day t?

t	$P(t)$

How many days will it take for the bacteria population to reach 8?

How many days will it take for the bacteria population to reach 16?

Roughly how long will it take for the population to reach 10?

We already know from our previous discussion that if $2^d = 10$, then $3 < d < 4$, and the table confirms that. At this point, we have an underestimate of 3 and an overestimate of 4 for d. How can we find better under and over estimates for d?

t	$P(t)$

EUREKA MATH™

©2015 Great Minds. eureka-math.org
ALG II-M3-SE-B2-1.3.0-08.2015

From our table, we now know another set of under and over estimates for the number d that we seek. What are they?

Continue this process of "squeezing" the number d between two numbers until you are confident you know the value of d to two decimal places.

t	$P(t)$

t	$P(t)$

What if we had wanted to find d to 5 decimal places?

To the nearest minute, when does the population of bacteria become 10?

t	$P(t)$

t	$P(t)$

t	$P(t)$

t	$P(t)$

Exercise

Use the method from the Example to approximate the solution to the equations below to two decimal places.

a. $2^x = 1,000$

b. $3^x = 1,000$

c. $4^x = 1,000$

d. $5^x = 1,000$

e. $6^x = 1,000$

f. $7^x = 1,000$

g. $8^x = 1,000$

h. $9^x = 1,000$

i. $11^x = 1,000$

j. $12^x = 1,000$

k. $13^x = 1,000$

l. $14^x = 1,000$

m. $15^x = 1,000$

n. $16^x = 1,000$

Problem Set

1. Solve each of the following equations for x using the same technique as was used in the Opening Exercise.

   a. $2^x = 32$

   b. $2^{x-3} = 2^{2x+5}$

   c. $2^{x^2-3x} = 2^{-2}$

   d. $2^x - 2^{4x-3} = 0$

   e. $2^{3x} \cdot 2^5 = 2^7$

   f. $2^{x^2-16} = 1$

   g. $3^{2x} = 27$

   h. $3^{\frac{2}{x}} = 81$

   i. $\dfrac{3^{x^2}}{3^{5x}} = 3^6$

2. Solve the equation $\dfrac{2^{2x}}{2^{x+5}} = 1$ algebraically using two different initial steps as directed below.

   a. Write each side as a power of 2.

   b. Multiply both sides by 2^{x+5}.

3. Find consecutive integers that are under and over estimates of the solutions to the following exponential equations.

   a. $2^x = 20$

   b. $2^x = 100$

   c. $3^x = 50$

   d. $10^x = 432{,}901$

   e. $2^{x-2} = 750$

   f. $2^x = 1.35$

4. Complete the following table to approximate the solution to $10^x = 34{,}198$ to three decimal places.

x	10^x	x	10^x	x	10^x	x	10^x
1	10	4.1		4.51		4.531	
2	100	4.2		4.52		4.532	
3	1,000	4.3		4.53		4.533	
4	10,000	4.4		4.54		4.534	
5	100,000	4.5				4.535	
		4.6					

5. Complete the following table to approximate the solution to $2^x = 19$ to three decimal places.

x	2^x	x	2^x	x	2^x	x	2^x

6. Approximate the solution to $5^x = 5,555$ to four decimal places.

7. A dangerous bacterial compound forms in a closed environment but is immediately detected. An initial detection reading suggests the concentration of bacteria in the closed environment is one percent of the fatal exposure level. This bacteria is known to double in concentration in a closed environment every hour and can be modeled by the function $P(t) = 100 \cdot 2^t$, where t is measured in hours.

    a. In the function $P(t) = 100 \cdot 2^t$, what does the 100 mean? What does the 2 mean?

    b. Doctors and toxicology professionals estimate that exposure to two-thirds of the bacteria's fatal concentration level will begin to cause sickness. Without consulting a calculator or other technology, offer a rough time limit for the inhabitants of the infected environment to evacuate in order to avoid sickness.

    c. A more conservative approach is to evacuate the infected environment before bacteria concentration levels reach one-third of fatal levels. Without consulting a calculator or other technology, offer a rough time limit for evacuation in this circumstance.

d. Use the method of the example to approximate when the bacteria concentration will reach 100% of the fatal exposure level, to the nearest minute.

t	2^t		t	2^t		t	2^t		t	2^t		t	2^t

This page intentionally left blank

Lesson 8: The "WhatPower" Function

Classwork

Opening Exercise

a. Evaluate each expression. The first two have been completed for you.

    i. $\text{WhatPower}_2(8) = 3$

    ii. $\text{WhatPower}_3(9) = 2$

    iii. $\text{WhatPower}_6(36) = \underline{\hspace{1.5cm}}$

    iv. $\text{WhatPower}_2(32) = \underline{\hspace{1.5cm}}$

    v. $\text{WhatPower}_{10}(1000) = \underline{\hspace{1.5cm}}$

    vi. $\text{WhatPower}_{10}(1000\ 000) = \underline{\hspace{1.5cm}}$

    vii. $\text{WhatPower}_{100}(1000\ 000) = \underline{\hspace{1.5cm}}$

    viii. $\text{WhatPower}_4(64) = \underline{\hspace{1.5cm}}$

    ix. $\text{WhatPower}_2(64) = \underline{\hspace{1.5cm}}$

    x. $\text{WhatPower}_9(3) = \underline{\hspace{1.5cm}}$

    xi. $\text{WhatPower}_5(\sqrt{5}) = \underline{\hspace{1.5cm}}$

    xii. $\text{WhatPower}_{\frac{1}{2}}\left(\frac{1}{8}\right) = \underline{\hspace{1.5cm}}$

    xiii. $\text{WhatPower}_{42}(1) = \underline{\hspace{1.5cm}}$

    xiv. $\text{WhatPower}_{100}(0.01) = \underline{\hspace{1.5cm}}$

    xv. $\text{WhatPower}_2\left(\frac{1}{4}\right) = \underline{\hspace{1.5cm}}$

        xvi. $\text{WhatPower}_{\frac{1}{4}}(2) = $ _____

b.   With your group members, write a definition for the function WhatPower_b, where b is a number.

Exercises 1–9

Evaluate the following expressions, and justify your answers.

1. $\text{WhatPower}_7(49)$

2. $\text{WhatPower}_0(7)$

3. $\text{WhatPower}_5(1)$

4. $\text{WhatPower}_1(5)$

5. $\text{WhatPower}_{-2}(16)$

6. $\text{WhatPower}_{-2}(32)$

7. $\text{WhatPower}_{\frac{1}{3}}(9)$

8. $\text{WhatPower}_{-\frac{1}{3}}(27)$

EUREKA MATH™

9. Describe the allowable values of b in the expression $\text{WhatPower}_b(x)$. When can we define a function $f(x) = \text{WhatPower}_b(x)$? Explain how you know.

Examples

1. $\log_2(8) = 3$

2. $\log_3(9) = 2$

3. $\log_6(36) = \underline{\hspace{1cm}}$

4. $\log_2(32) = \underline{\hspace{1cm}}$

5. $\log_{10}(1000) = \underline{\hspace{1cm}}$

6. $\log_{42}(1) = \underline{\hspace{1cm}}$

7. $\log_{100}(0.01) = \underline{\hspace{1cm}}$

8. $\log_2\left(\frac{1}{4}\right) = \underline{\hspace{1cm}}$

Exercise 10

10. Compute the value of each logarithm. Verify your answers using an exponential statement.

    a. $\log_2(32)$

b. $\log_3(81)$

c. $\log_9(81)$

d. $\log_5(625)$

e. $\log_{10}(1000000000)$

f. $\log_{1000}(1000000000)$

g. $\log_{13}(13)$

h. $\log_{13}(1)$

i. $\log_7(\sqrt{7})$

j. $\log_9(27)$

k. $\log_{\sqrt{7}}(7)$

l. $\log_{\sqrt{7}}\left(\dfrac{1}{49}\right)$

m. $\log_x(x^2)$

EUREKA
MATH™

Lesson Summary

- If three numbers L, b, and x are related by $x = b^L$, then L is the *logarithm base b of x*, and we write $\log_b(x) = L$. That is, the value of the expression $\log_b(x)$ is the power of b needed to obtain x.
- Valid values of b as a base for a logarithm are $0 < b < 1$ and $b > 1$.

Problem Set

1. Rewrite each of the following in the form $\text{WhatPower}_b(x) = L$.

   a. $3^5 = 243$ b. $6^{-3} = \dfrac{1}{216}$ c. $9^0 = 1$

2. Rewrite each of the following in the form $\log_b(x) = L$.

   a. $16^{\frac{1}{4}} = 2$ b. $10^3 = 1000$ c. $b^k = r$

3. Rewrite each of the following in the form $b^L = x$.

   a. $\log_5(625) = 4$ b. $\log_{10}(0.1) = -1$ c. $\log_{27} 9 = \dfrac{2}{3}$

4. Consider the logarithms base 2. For each logarithmic expression below, either calculate the value of the expression or explain why the expression does not make sense.

   a. $\log_2(1024)$

   b. $\log_2(128)$

   c. $\log_2(\sqrt{8})$

   d. $\log_2\left(\dfrac{1}{16}\right)$

   e. $\log_2(0)$

   f. $\log_2\left(-\dfrac{1}{32}\right)$

5. Consider the logarithms base 3. For each logarithmic expression below, either calculate the value of the expression or explain why the expression does not make sense.

   a. $\log_3(243)$

   b. $\log_3(27)$

   c. $\log_3(1)$

   d. $\log_3\left(\dfrac{1}{3}\right)$

   e. $\log_3(0)$

   f. $\log_3\left(-\dfrac{1}{3}\right)$

6. Consider the logarithms base 5. For each logarithmic expression below, either calculate the value of the expression or explain why the expression does not make sense.

   a. $\log_5(3125)$

   b. $\log_5(25)$

   c. $\log_5(1)$

   d. $\log_5\left(\dfrac{1}{25}\right)$

   e. $\log_5(0)$

   f. $\log_5\left(-\dfrac{1}{25}\right)$

7. Is there any positive number b so that the expression $\log_b(0)$ makes sense? Explain how you know.

8. Is there any positive number b so that the expression $\log_b(-1)$ makes sense? Explain how you know.

9. Verify each of the following by evaluating the logarithms.

   a. $\log_2(8) + \log_2(4) = \log_2(32)$

   b. $\log_3(9) + \log_3(9) = \log_3(81)$

   c. $\log_4(4) + \log_4(16) = \log_4(64)$

   d. $\log_{10}(10^3) + \log_{10}(10^4) = \log_{10}(10^7)$

10. Looking at the results from Problem 9, do you notice a trend or pattern? Can you make a general statement about the value of $\log_b(x) + \log_b(y)$?

11. To evaluate $\log_2(3)$, Autumn reasoned that since $\log_2(2) = 1$ and $\log_2(4) = 2$, $\log_2(3)$ must be the average of 1 and 2 and therefore $\log_2(3) = 1.5$. Use the definition of logarithm to show that $\log_2(3)$ cannot be 1.5. Why is her thinking not valid?

12. Find the value of each of the following.

   a. If $x = \log_2(8)$ and $y = 2^x$, find the value of y.

   b. If $\log_2(x) = 6$, find the value of x.

   c. If $r = 2^6$ and $s = \log_2(r)$, find the value of s.

EUREKA
MATH™

Lesson 9: Logarithms—How Many Digits Do You Need?

Classwork

Opening Exercise

a. Evaluate WhatPower$_2$(8). State your answer as a logarithm, and evaluate it.

b. Evaluate WhatPower$_5$(625). State your answer as a logarithm, and evaluate it.

Exploratory Challenge

Autumn is starting a new club with eight members including herself. She wants everyone to have a secret identification code made up of only A's and B's, known as an ID code. For example, using two characters, her ID code could be AA.

a. Using A's and B's, can Autumn assign each club member a unique two-character ID code using only A's and B's? Justify your answer. Here's what Autumn has so far.

Club Member Name	Secret ID Code
Autumn	AA
Kris	
Tia	
Jimmy	

Club Member Name	Secret ID Code
Robert	
Jillian	
Benjamin	
Scott	

b. Using A's and B's, how many characters would be needed to assign each club member a unique ID code? Justify your answer by showing the ID codes you would assign to each club member by completing the table above (adjust Autumn's ID if needed).

When the club grew to 16 members, Autumn started noticing a pattern.

Using A's and B's:

    i.   Two people could be given a secret ID code with 1 letter: A and B.

    ii.   Four people could be given a secret ID code with 2 letters: AA, BA, AB, BB.

    iii.   Eight people could be given a secret ID code with 3 letters: AAA, BAA, ABA, BBA, AAB, BAB, ABB, BBB.

c.   Complete the following statement, and list the secret ID codes for the 16 people:

16 people could be given a secret ID code with _____ letters using A's and B's.

Club Member Name	Secret ID Code
Autumn	
Kris	
Tia	
Jimmy	
Robert	
Jillian	
Benjamin	
Scott	

Club Member Name	Secret ID Code
Gwen	
Jerrod	
Mykel	
Janette	
Nellie	
Serena	
Ricky	
Mia	

d.   Describe the pattern in words. What type of function could be used to model this pattern?

EUREKA
MATH™

©2015 Great Minds. eureka-math.org
ALG II-M3-SE-B2-1.3.0-08.2015

Exercises 1–2

In the previous problems, the letters A and B were like the digits in a number. A four-digit ID code for Autumn's club could be any four-letter arrangement of A's and B's because in her ID system, the only digits are the letters A and B.

1. When Autumn's club grows to include more than 16 people, she will need five digits to assign a unique ID code to each club member. What is the maximum number of people that could be in the club before she needs to switch to a six-digit ID code? Explain your reasoning.

2. If Autumn has 256 members in her club, how many digits would she need to assign each club member a unique ID code using only A's and B's? Show how you got your answers.

Example

A thousand people are given unique identifiers made up of the digits 0, 1, 2, …, 9. How many digits would be needed for each ID number?

Exercises 3–4

3. There are approximately 317 million people in the United States. Compute and use $\log(100000000)$ and $\log(1000000\,000)$ to explain why Social Security numbers are 9 digits long.

4. There are many more telephones than the number of people in the United States because of people having home phones, cell phones, business phones, fax numbers, etc. Assuming we need at most 10 billion phone numbers in the United States, how many digits would be needed so that each phone number is unique? Is this reasonable? Explain.

Problem Set

1. The student body president needs to assign each officially sanctioned club on campus a unique ID code for purposes of tracking expenses and activities. She decides to use the letters A, B, and C to create a unique three-character code for each club.

   a. How many clubs can be assigned a unique ID code according to this proposal?

   b. There are actually over 500 clubs on campus. Assuming the student body president still wants to use the letters A, B, and C, how many characters would be needed to generate a unique ID code for each club?

2. Can you use the numbers 1, 2, 3, and 4 in a combination of four digits to assign a unique ID code to each of 500 people? Explain your reasoning.

3. Automobile license plates typically have a combination of letters (26) and numbers (10). Over time, the state of New York has used different criteria to assign vehicle license plate numbers.

   a. From 1973 to 1986, the state used a 3-letter and 4-number code where the three letters indicated the county where the vehicle was registered. Essex County had 13 different 3-letter codes in use. How many cars could be registered to this county?

   b. Since 2001, the state has used a 3-letter and 4-number code but no longer assigns letters by county. Is this coding scheme enough to register 10 million vehicles?

4. The Richter scale uses base 10 logarithms to assign a magnitude to an earthquake based on the amount of force released at the earthquake's source as measured by seismographs in various locations.

   a. Explain the difference between an earthquake that is assigned a magnitude of 5 versus one assigned a magnitude of 7.

   b. An earthquake with magnitude 2 can usually only be felt by people located near the earthquake's origin, called its *epicenter*. The strongest earthquake on record occurred in Chile in 1960 with magnitude 9.5. How many times stronger is the force of an earthquake with magnitude 9.5 than the force of an earthquake with magnitude 2?

   c. What is the magnitude of an earthquake whose force is 1,000 times greater than a magnitude 4.3 quake?

5. Sound pressure level is measured in decibels (dB) according to the formula $L = 10 \log\left(\frac{I}{I_0}\right)$, where I is the intensity of the sound and I_0 is a reference intensity that corresponds to a barely perceptible sound.

   a. Explain why this formula would assign 0 decibels to a barely perceptible sound.

   b. Decibel levels above 120 dB can be painful to humans. What would be the intensity that corresponds to this level?

This page intentionally left blank

Lesson 10: Building Logarithmic Tables

Classwork

Opening Exercise

Find the value of the following expressions without using a calculator.

$\text{WhatPower}_{10}(1000)$ $\log_{10}(1000)$

$\text{WhatPower}_{10}(100)$ $\log_{10}(100)$

$\text{WhatPower}_{10}(10)$ $\log_{10}(10)$

$\text{WhatPower}_{10}(1)$ $\log_{10}(1)$

$\text{WhatPower}_{10}\left(\dfrac{1}{10}\right)$ $\log_{10}\left(\dfrac{1}{10}\right)$

$\text{WhatPower}_{10}\left(\dfrac{1}{100}\right)$ $\log_{10}\left(\dfrac{1}{100}\right)$

Formulate a rule based on your results above: If k is an integer, then $\log_{10}(10^k) = $ _____.

Example 1

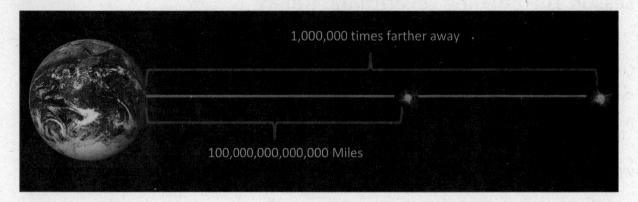

Exercises

1. Find two consecutive powers of 10 so that 30 is between them. That is, find an integer exponent k so that $10^k < 30 < 10^{k+1}$.

2. From your result in Exercise 1, $\log(30)$ is between which two integers?

3. Find a number k to one decimal place so that $10^k < 30 < 10^{k+0.1}$, and use that to find under and over estimates for $\log(30)$.

4. Find a number k to two decimal places so that $10^k < 30 < 10^{k+0.01}$, and use that to find under and over estimates for $\log(30)$.

Lesson 10:   Building Logarithmic Tables

©2015 Great Minds. eureka-math.org
ALG II-M3-SE-B2-1.3.0-08.2015

5. Repeat this process to approximate the value of log(30) to 4 decimal places.

6. Verify your result on your calculator, using the $\boxed{\text{LOG}}$ button.

7. Use your calculator to complete the following table. Round the logarithms to 4 decimal places.

x	$\log(x)$		x	$\log(x)$		x	$\log(x)$
1			10			100	
2			20			200	
3			30			300	
4			40			400	
5			50			500	
6			60			600	
7			70			700	
8			80			800	
9			90			900	

8. What pattern(s) can you see in the table from Exercise 7 as x is multiplied by 10? Write the pattern(s) using logarithmic notation.

9. What pattern would you expect to find for $\log(1000x)$? Make a conjecture, and test it to see whether or not it appears to be valid.

10. Use your results from Exercises 8 and 9 to make a conjecture about the value of $\log(10^k \cdot x)$ for any positive integer k.

11. Use your calculator to complete the following table. Round the logarithms to 4 decimal places.

x	$\log(x)$	x	$\log(x)$	x	$\log(x)$
1		0.1		0.01	
2		0.2		0.02	
3		0.3		0.03	
4		0.4		0.04	
5		0.5		0.05	
6		0.6		0.06	
7		0.7		0.07	
8		0.8		0.08	
9		0.9		0.09	

12. What pattern(s) can you see in the table from Exercise 11? Write them using logarithmic notation.

Lesson 10:   Building Logarithmic Tables

©2015 Great Minds. eureka-math.org
ALG II-M3-SE-B2-1.3.0-08.2015

EUREKA MATH™

13. What pattern would you expect to find for $\log\left(\frac{x}{1000}\right)$? Make a conjecture, and test it to see whether or not it appears to be valid.

14. Combine your results from Exercises 10 and 12 to make a conjecture about the value of the logarithm for a multiple of a power of 10; that is, find a formula for $\log(10^k \cdot x)$ for any integer k.

Lesson Summary

- The notation $\log(x)$ is used to represent $\log_{10}(x)$.
- For integers k, $\log(10^k) = k$.
- For integers m and n, $\log(10^m \cdot 10^n) = \log(10^m) + \log(10^n)$.
- For integers k and positive real numbers x, $\log(10^k \cdot x) = k + \log(x)$.

Problem Set

1. Complete the following table of logarithms without using a calculator; then, answer the questions that follow.

x	$\log(x)$
1,000,000	
100,000	
10,000	
1000	
100	
10	

x	$\log(x)$
0.1	
0.01	
0.001	
0.0001	
0.00001	
0.000001	

   a. What is $\log(1)$? How does that follow from the definition of a base-10 logarithm?

   b. What is $\log(10^k)$ for an integer k? How does that follow from the definition of a base-10 logarithm?

   c. What happens to the value of $\log(x)$ as x gets really large?

   d. For $x > 0$, what happens to the value of $\log(x)$ as x gets really close to zero?

2. Use the table of logarithms below to estimate the values of the logarithms in parts (a)–(h).

x	$\log(x)$
2	0.3010
3	0.4771
5	0.6990
7	0.8451
11	1.0414
13	1.1139

   a. $\log(70\ 000)$

   b. $\log(0.0011)$

   c. $\log(20)$

   d. $\log(0.00005)$

   e. $\log(130\ 000)$

   f. $\log(3000)$

   g. $\log(0.07)$

   h. $\log(11000000)$

---

3. If $\log(n) = 0.6$, find the value of $\log(10n)$.

4. If m is a positive integer and $\log(m) \approx 3.8$, how many digits are there in m? Explain how you know.

5. If m is a positive integer and $\log(m) \approx 9.6$, how many digits are there in m? Explain how you know.

6. Vivian says $\log(452\,000) = 5 + \log(4.52)$, while her sister Lillian says that $\log(452\,000) = 6 + \log(0.452)$. Which sister is correct? Explain how you know.

7. Write the base-10 logarithm of each number in the form $k + \log(x)$, where k is the exponent from the scientific notation, and x is a positive real number.

    a. 2.4902×10^4

    b. 2.58×10^{13}

    c. 9.109×10^{-31}

8. For each of the following statements, write the number in scientific notation, and then write the logarithm base 10 of that number in the form $k + \log(x)$, where k is the exponent from the scientific notation, and x is a positive real number.

    a. The speed of sound is 1116 ft/s.

    b. The distance from Earth to the sun is 93 million miles.

    c. The speed of light is $29,980,000,000$ cm/s.

    d. The weight of the earth is $5,972,000,000,000\,000,000,000,000$ kg.

    e. The diameter of the nucleus of a hydrogen atom is 0.00000000000000175 m.

    f. For each part (a)–(e), you have written each logarithm in the form $k + \log(x)$, for integers k and positive real numbers x. Use a calculator to find the values of the expressions $\log(x)$. Why are all of these values between 0 and 1?

This page intentionally left blank

Lesson 11: The Most Important Property of Logarithms

Classwork

Opening Exercise

Use the logarithm table below to calculate the specified logarithms.

x	$\log(x)$
1	0
2	0.3010
3	0.4771
4	0.6021
5	0.6990
6	0.7782
7	0.8451
8	0.9031
9	0.9542

a. $\log(80)$

b. $\log(7000)$

c. $\log(0.00006)$

d. $\log(3.0 \times 10^{27})$

e. $\log(9.0 \times 10^{k})$ for an integer k

Exercises 1–5

1. Use your calculator to complete the following table. Round the logarithms to four decimal places.

x	$\log(x)$		x	$\log(x)$
1	0		10	
2	0.3010		12	
3	0.4771		16	
4	0.6021		18	
5	0.6990		20	
6	0.7782		25	
7	0.8451		30	
8	0.9031		36	
9	0.9542		100	

2. Calculate the following values. Do they appear anywhere else in the table?

a. $\log(2) + \log(4)$

b. $\log(2) + \log(6)$

c. $\log(3) + \log(4)$

d. $\log(6) + \log(6)$

e. $\log(2) + \log(18)$

f. $\log(3) + \log(12)$

EUREKA MATH™

3. What pattern(s) can you see in Exercise 2 and the table from Exercise 1? Write them using logarithmic notation.

4. What pattern would you expect to find for $\log(x^2)$? Make a conjecture, and test it to see whether or not it appears to be valid.

5. Make a conjecture for a logarithm of the form $\log(xyz)$, where x, y, and z are positive real numbers. Provide evidence that your conjecture is valid.

Example 1

Use the logarithm table from Exercise 1 to approximate the following logarithms.

    a. $\log(14)$

    b. $\log(35)$

c. $\log(72)$

d. $\log(121)$

Exercises 6–8

6. Use your calculator to complete the following table. Round the logarithms to four decimal places.

x	$\log(x)$
2	
4	
5	
8	
10	
16	
20	
50	
100	

x	$\log(x)$
$\dfrac{1}{2}$	
$\dfrac{1}{4}$	
$\dfrac{1}{5}$	
$\dfrac{1}{8}$	
$\dfrac{1}{10}$	
$\dfrac{1}{16}$	
$\dfrac{1}{20}$	
$\dfrac{1}{50}$	
$\dfrac{1}{100}$	

7. What pattern(s) can you see in the table from Exercise 6? Write a conjecture using logarithmic notation.

8. Use the definition of logarithm to justify the conjecture you found in Exercise 7.

©2015 Great Minds. eureka-math.org
ALG II-M3-SE-B2-1.3.0-08.2015

Example 2

Use the logarithm tables and the rules we have discovered to estimate the following logarithms to four decimal places.

    a. $\log(2100)$

    b. $\log(0.00049)$

    c. $\log(42000000)$

    d. $\log\left(\frac{1}{640}\right)$

Lesson Summary

- The notation $\log(x)$ is used to represent $\log_{10}(x)$.

- The most important property of base 10 logarithms is that for positive real numbers x and y,

$$\log(xy) = \log(x) + \log(y).$$

- For positive real numbers x,

$$\log\left(\frac{1}{x}\right) = -\log(x).$$

Problem Set

1. Use the table of logarithms to the right to estimate the value of the logarithms in parts (a)–(t).

   a. $\log(25)$

   b. $\log(27)$

   c. $\log(33)$

   d. $\log(55)$

   e. $\log(63)$

   f. $\log(75)$

   g. $\log(81)$

   h. $\log(99)$

   i. $\log(350)$

   j. $\log(0.0014)$

   k. $\log(0.077)$

   l. $\log(49000)$

   m. $\log(1.69)$

   n. $\log(6.5)$

   o. $\log\left(\frac{1}{30}\right)$

   p. $\log\left(\frac{1}{35}\right)$

   q. $\log\left(\frac{1}{40}\right)$

   r. $\log\left(\frac{1}{42}\right)$

   s. $\log\left(\frac{1}{50}\right)$

   t. $\log\left(\frac{1}{64}\right)$

x	$\log(x)$
2	0.30
3	0.48
5	0.70
7	0.85
11	1.04
13	1.11

Lesson 11:   The Most Important Property of Logarithms

EUREKA MATH™

2. Reduce each expression to a single logarithm of the form $\log(x)$.

   a. $\log(5) + \log(7)$

   b. $\log(3) + \log(9)$

   c. $\log(15) - \log(5)$

   d. $\log(8) + \log\left(\frac{1}{4}\right)$

3. Use properties of logarithms to write the following expressions involving logarithms of only prime numbers:

   a. $\log(2500)$

   b. $\log(0.00063)$

   c. $\log(1250)$

   d. $\log(26000000)$

4. Use properties of logarithms to show that $\log(2) - \log\left(\frac{1}{13}\right) = \log(26)$.

5. Use properties of logarithms to show that $\log(3) + \log(4) + \log(5) - \log(6) = 1$.

6. Use properties of logarithms to show that $\log\left(\frac{1}{2} - \frac{1}{3}\right) + \log(2) = -\log(3)$.

7. Use properties of logarithms to show that $\log\left(\frac{1}{3} - \frac{1}{4}\right) + \left(\log\left(\frac{1}{3}\right) - \log\left(\frac{1}{4}\right)\right) = -2\log(3)$.

This page intentionally left blank

Lesson 12: Properties of Logarithms

Classwork

Opening Exercise

Use the approximation $\log(2) \approx 0.3010$ to approximate the values of each of the following logarithmic expressions.

a. $\log(20)$

b. $\log(0.2)$

c. $\log(2^4)$

Exercises

For Exercises 1–6, explain why each statement below is a property of base-10 logarithms.

1. Property 1: $\log(1) = 0$.

2. Property 2: $\log(10) = 1$.

3. Property 3: For all real numbers r, $\log(10^r) = r$.

4. Property 4: For any $x > 0$, $10^{\log(x)} = x$.

EUREKA
MATH™

5. **Property 5:** For any positive real numbers x and y, $\log(x \cdot y) = \log(x) + \log(y)$.
   Hint: Use an exponent rule as well as Property 4.

6. **Property 6:** For any positive real number x and any real number r, $\log(x^r) = r \cdot \log(x)$.
   Hint: Use an exponent rule as well as Property 4.

7. Apply properties of logarithms to rewrite the following expressions as a single logarithm or number.

   a. $\frac{1}{2}\log(25) + \log(4)$

   b. $\frac{1}{3}\log(8) + \log(16)$

   c. $3\log(5) + \log(0.8)$

8. Apply properties of logarithms to rewrite each expression as a sum of terms involving numbers, $\log(x)$, and $\log(y)$, where x and y are positive real numbers.

   a. $\log(3x^2y^5)$

   b. $\log(\sqrt{x^7y^3})$

9. In mathematical terminology, logarithms are *well defined* because if $X = Y$, then $\log(X) = \log(Y)$ for $X, Y > 0$. This means that if you want to solve an equation involving exponents, you can apply a logarithm to both sides of the equation, just as you can take the square root of both sides when solving a quadratic equation. You do need to be careful not to take the logarithm of a negative number or zero.

Use the property stated above to solve the following equations.

a. $10^{10x} = 100$

b. $10^{x-1} = \dfrac{1}{10^{x+1}}$

c. $100^{2x} = 10^{3x-1}$

10. Solve the following equations.

a. $10^x = 2^7$

b. $10^{x^2+1} = 15$

c. $4^x = 5^3$

EUREKA
MATH

Lesson 12:   Properties of Logarithms

S.85

©2015 Great Minds. eureka-math.org
ALG II-M3-SE-B2-1.3.0-08.2015

Lesson Summary

We have established the following properties for base 10 logarithms, where x and y are positive real numbers and r is any real number:

1. $\log(1) = 0$
2. $\log(10) = 1$
3. $\log(10^r) = r$
4. $10^{\log(x)} = x$
5. $\log(x \cdot y) = \log(x) + \log(y)$
6. $\log(x^r) = r \cdot \log(x)$

Additional properties not yet established are the following:

7. $\log\left(\frac{1}{x}\right) = -\log(x)$
8. $\log\left(\frac{x}{y}\right) = \log(x) - \log(y)$

Also, logarithms are well defined, meaning that for $X, Y > 0$, if $X = Y$, then $\log(X) = \log(Y)$.

Problem Set

1. Use the approximate logarithm values below to estimate the value of each of the following logarithms. Indicate which properties you used.

$$\log(2) = 0.3010 \qquad \log(3) = 0.4771$$
$$\log(5) = 0.6990 \qquad \log(7) = 0.8451$$

   a. $\log(6)$

   b. $\log(15)$

   c. $\log(12)$

   d. $\log(10^7)$

   e. $\log\left(\frac{1}{5}\right)$

   f. $\log\left(\frac{3}{7}\right)$

   g. $\log(\sqrt[4]{2})$

Lesson 12:   Properties of Logarithms

©2015 Great Minds. eureka-math.org
ALG II-M3-SE-B2-1.3.0-08.2015

2. Let $\log(X) = r$, $\log(Y) = s$, and $\log(Z) = t$. Express each of the following in terms of r, s, and t.

   a. $\log\left(\frac{X}{Y}\right)$

   b. $\log(YZ)$

   c. $\log(X^r)$

   d. $\log(\sqrt[3]{Z})$

   e. $\log\left(\sqrt[4]{\dfrac{Y}{Z}}\right)$

   f. $\log(XY^2Z^3)$

3. Use the properties of logarithms to rewrite each expression in an equivalent form containing a single logarithm.

   a. $\log\left(\frac{13}{5}\right) + \log\left(\frac{5}{4}\right)$

   b. $\log\left(\frac{5}{6}\right) - \log\left(\frac{2}{3}\right)$

   c. $\frac{1}{2}\log(16) + \log(3) + \log\left(\frac{1}{4}\right)$

4. Use the properties of logarithms to rewrite each expression in an equivalent form containing a single logarithm.

   a. $\log(\sqrt{x}) + \frac{1}{2}\log\left(\frac{1}{x}\right) + 2\log(x)$

   b. $\log(\sqrt[5]{x}) + \log(\sqrt[5]{x^4})$

   c. $\log(x) + 2\log(y) - \frac{1}{2}\log(z)$

   d. $\frac{1}{3}\left(\log(x) - 3\log(y) + \log(z)\right)$

   e. $2(\log(x) - \log(3y)) + 3(\log(z) - 2\log(x))$

5. In each of the following expressions, x, y, and z represent positive real numbers. Use properties of logarithms to rewrite each expression in an equivalent form containing only $\log(x)$, $\log(y)$, $\log(z)$, and numbers.

   a. $\log\left(\dfrac{3x^2y^4}{\sqrt{z}}\right)$

   b. $\log\left(\dfrac{42\sqrt[3]{xy^7}}{x^2z}\right)$

   c. $\log\left(\dfrac{100x^2}{y^3}\right)$

   d. $\log\left(\sqrt{\dfrac{x^3y^2}{10z}}\right)$

   e. $\log\left(\dfrac{1}{10x^2z}\right)$

6. Express $\log\left(\frac{1}{x} - \frac{1}{x+1}\right) + \left(\log\left(\frac{1}{x}\right) - \log\left(\frac{1}{x+1}\right)\right)$ as a single logarithm for positive numbers x.

7. Show that $\log\left(x + \sqrt{x^2 - 1}\right) + \log\left(x - \sqrt{x^2 - 1}\right) = 0$ for $x \geq 1$.

8. If $xy = 10^{3.67}$ for some positive real numbers x and y, find the value of $\log(x) + \log(y)$.

9. Solve the following exponential equations by taking the logarithm base 10 of both sides. Leave your answers stated in terms of logarithmic expressions.

   a. $10^{x^2} = 320$

   b. $10^{\frac{x}{8}} = 300$

   c. $10^{3x} = 400$

   d. $5^{2x} = 200$

   e. $3^x = 7^{-3x+2}$

10. Solve the following exponential equations.

    a. $10^x = 3$

    b. $10^y = 30$

    c. $10^z = 300$

    d. Use the properties of logarithms to justify why x, y, and z form an arithmetic sequence whose constant difference is 1.

11. Without using a calculator, explain why the solution to each equation must be a real number between 1 and 2.

    a. $11^x = 12$

    b. $21^x = 30$

    c. $100^x = 2000$

    d. $\left(\frac{1}{11}\right)^x = 0.01$

    e. $\left(\frac{2}{3}\right)^x = \frac{1}{2}$

    f. $99^x = 9000$

EUREKA MATH™

12. Express the exact solution to each equation as a base-10 logarithm. Use a calculator to approximate the solution to the nearest 1000th.

   a. $11^x = 12$

   b. $21^x = 30$

   c. $100^x = 2000$

   d. $\left(\frac{1}{11}\right)^x = 0.01$

   e. $\left(\frac{2}{3}\right)^x = \frac{1}{2}$

   f. $99^x = 9000$

13. Show that for any real number r, the solution to the equation $10^x = 3 \cdot 10^r$ is $\log(3) + r$.

14. Solve each equation. If there is no solution, explain why.

   a. $3 \cdot 5^x = 21$

   b. $10^{x-3} = 25$

   c. $10^x + 10^{x+1} = 11$

   d. $8 - 2^x = 10$

15. Solve the following equation for n: $A = P(1 + r)^n$.

16. In this exercise, we will establish a formula for the logarithm of a sum. Let $L = \log(x + y)$, where $x, y > 0$.

   a. Show $\log(x) + \log\left(1 + \frac{y}{x}\right) = L$. State as a property of logarithms after showing this is a true statement.

   b. Use part (a) and the fact that $\log(100) = 2$ to rewrite $\log(365)$ as a sum.

   c. Rewrite 365 in scientific notation, and use properties of logarithms to express $\log(365)$ as a sum of an integer and a logarithm of a number between 0 and 10.

   d. What do you notice about your answers to (b) and (c)?

   e. Find two integers that are upper and lower estimates of $\log(365)$.

This page intentionally left blank

Lesson 13: Changing the Base

Classwork

Exercises

1. Assume that x, a, and b are all positive real numbers, so that $a \neq 1$ and $b \neq 1$. What is $\log_b(x)$ in terms of $\log_a(x)$? The resulting equation allows us to change the base of a logarithm from a to b.

2. Approximate each of the following logarithms to four decimal places. Use the $\boxed{\text{LOG}}$ key on your calculator rather than logarithm tables, first changing the base of the logarithm to 10 if necessary.

   a. $\log(3^2)$

   b. $\log_3(3^2)$

   c. $\log_2(3^2)$

©2015 Great Minds. eureka-math.org
ALG II-M3-SE-B2-1.3.0-08.2015

3. In Lesson 12, we justified a number of properties of base-10 logarithms. Working in pairs, justify the following properties of base b logarithms:

a. $\log_b(1) = 0$

b. $\log_b(b) = 1$

c. $\log_b(b^r) = r$

d. $b^{\log_b(x)} = x$

e. $\log_b(x \cdot y) = \log_b(x) + \log_b(y)$

EUREKA
MATH™

f. $\log_b(x^r) = r \cdot \log_b(x)$

g. $\log_b\left(\dfrac{1}{x}\right) = -\log_b(x)$

h. $\log_b\left(\dfrac{x}{y}\right) = \log_b(x) - \log_b(y)$

4.   Use the $\boxed{\text{LN}}$ and $\boxed{\text{LOG}}$ keys on your calculator to find the value of each logarithm to four decimal places.

a. $\ln(1)$ $\log(1)$

b. $\ln(3)$ $\log(3)$

c. $\ln(10)$ $\log(10)$

d. $\ln(25)$ $\log(25)$

e. $\ln(100)$ $\log(100)$

5.   Make a conjecture that compares values of $\log(x)$ to $\ln(x)$ for $x \geq 1$.

6. Justify your conjecture in Exercise 5 using the change of base formula.

7. Write as a single logarithm.

    a. $\ln(4) - 3\ln\left(\frac{1}{3}\right) + \ln(2)$

    b. $\ln(5) + \frac{3}{5}\ln(32) - \ln(4)$

8. Write each expression as a sum or difference of constants and logarithms of simpler terms.

    a. $\ln\left(\frac{\sqrt{5x^3}}{e^2}\right)$

    b. $\ln\left(\frac{(x+y)^2}{x^2+y^2}\right)$

TORY OF FUNCTIONS
Lesson 13 M3

ALGEBRA II

A STORY OF FUNCTIONS
Lesson 13 M3

ALGEBRA II

olve each exponential equation.

$3^{2x} = 81$　　h. $2^x = 81$

$6^{3x} = 36^{x+1}$　　i. $8 = 3^x$

$625 = 5^{3x}$　　j. $6^{x+2} = 12$

$25^{4-x} = 5^{3x}$　　k. $10^{x+4} = 27$

$32^{x-1} = \frac{1}{2}$　　l. $2^{x+1} = 3^{1-x}$

m. $3^{2x-3} = 2^{x+4}$

$\frac{4^{2x}}{2^{x-3}} = 1$　　n. $e^{2x} = 5$

o. $e^{x-1} = 6$

$\frac{1}{8^{2x-4}} = 1$

Problem 9(e) of Lesson 12, you solved the equation $3^x = 7^{-3x+2}$ using the logarithm base 10.

Solve $3^x = 7^{-3x+2}$ using the logarithm base 3.

Apply the change of base formula to show that your answer to part (a) agrees with your answer to Problem 9(e) of Lesson 12.

Solve $3^x = 7^{-3x+2}$ using the logarithm base 7.

Apply the change of base formula to show that your answer to part (c) also agrees with your answer to Problem 9(e) of Lesson 12.

arl solved the equation $2^x = 10$ as follows:

$\log(2^x) = \log(10)$

$x \log(2) = 1$

$x = \frac{1}{\log(2)}.$

s solved the equation $2^x = 10$ as follows:

$\log_2(2^x) = \log_2(10)$

$x \log_2(2) = \log_2(10)$

$x = \log_2(10).$

earl correct? Is Jess correct? Explain how you know.

Lesson Summary

We have established a formula for changing the base of logarithms from b to a:

$$\log_b(x) = \frac{\log_a(x)}{\log_a(b)}.$$

In particular, the formula allows us to change logarithms base b to common or natural logarithms, which are the only two kinds of logarithms that most calculators compute:

$$\log_b(x) = \frac{\log(x)}{\log(b)} = \frac{\ln(x)}{\ln(b)}.$$

We have also established the following properties for base b logarithms. If x, y, a, and b are all positive real numbers with $a \neq 1$ and $b \neq 1$ and r is any real number, then:

1. $\log_b(1) = 0$
2. $\log_b(b) = 1$
3. $\log_b(b^r) = r$
4. $b^{\log_b(x)} = x$
5. $\log_b(x \cdot y) = \log_b(x) + \log_b(y)$
6. $\log_b(x^r) = r \cdot \log_b(x)$
7. $\log_b\left(\frac{1}{x}\right) = -\log_b(x)$
8. $\log_b\left(\frac{x}{y}\right) = \log_b(x) - \log_b(y)$

Problem Set

1. Evaluate each of the following logarithmic expressions, approximating to four decimal places if necessary. Use the LN or LOG key on your calculator rather than a table.

    a. $\log_8(16)$

    b. $\log_7(11)$

    c. $\log_3(2) + \log_2(3)$

2. Use logarithmic properties and the fact that $\ln(2) \approx 0.69$ and $\ln(3) \approx 1.10$ to approximate the value of each of the following logarithmic expressions. Do not use a calculator.

    a. $\ln(e^4)$

    b. $\ln(6)$

    c. $\ln(108)$

    d. $\ln\left(\frac{8}{3}\right)$

Lesson 13: Changing the Base

EUREKA MATH

©2015 Great Minds. eureka-math.org
ALG II-M3-SE-B2-1.3.0-08.2015

EUREKA MATH

Lesson 13: Changing the Base

S.95

©2015 Great Minds. eureka-math.org
ALG II-M3-SE-B2-1.3.0-08.2015

3. Compare the values of $\log_{\frac{1}{9}}(10)$ and $\log_9\left(\frac{1}{10}\right)$ without using a calculator.

4. Show that for any positive numbers a and b with $a \neq 1$ and $b \neq 1$, $\log_a(b) \cdot \log_b(a) = 1$.

5. Express x in terms of a, e, and y if $\ln(x) - \ln(y) = 2a$.

6. Rewrite each expression in an equivalent form that only contains one base 10 logarithm.

   a. $\log_2(800)$

   b. $\log_x\left(\frac{1}{10}\right)$, for positive real values of $x \neq 1$

   c. $\log_5(12500)$

   d. $\log_3(0.81)$

7. Write each number in terms of natural logarithms, and then use the properties of logarithms to show that it is a rational number.

   a. $\log_9\left(\sqrt{27}\right)$

   b. $\log_8(32)$

   c. $\log_4\left(\frac{1}{8}\right)$

8. Write each expression as an equivalent expression with a single logarithm. Assume x, y, and z are positive real numbers.

   a. $\ln(x) + 2\ln(y) - 3\ln(z)$

   b. $\frac{1}{2}\left(\ln(x+y) - \ln(z)\right)$

   c. $(x+y) + \ln(z)$

9. Rewrite each expression as sums and differences in terms of $\ln(x)$, $\ln(y)$, and $\ln(z)$.

   a. $\ln(xyz^3)$

   b. $\ln\left(\frac{e^3}{xyz}\right)$

   c. $\ln\left(\sqrt{\frac{x}{y}}\right)$

10. Use base-5 logarithms to rewrite each exponential equation as a logarithmic equation, and s
    equation. Use the change of base formula to convert to a base-10 logarithm that can be eva
    Give each answer to 4 decimal places. If an equation has no solution, explain why.

    a. $5^{2x} = 20$

    b. $75 = 10 \cdot 5^{x-1}$

    c. $5^{2+x} - 5^x = 10$

    d. $5^{x^2} = 0.25$

11. In Lesson 6, you discovered that $\log(x \cdot 10^k) = k + \log(x)$ by looking at a table of logarithm
    logarithms to justify this property for an arbitrary base $b > 0$ with $b \neq 1$. That is, show that
    $\log_b(x \cdot b^k) = k + \log_b(x)$.

12. Larissa argued that since $\log_2(2) = 1$ and $\log_2(4) = 2$, then it must be true that $\log_2(3) = $
    Explain how you know.

13. Extension: Suppose that there is some positive number b so that

    $$\log_b(2) = 0.36$$
    $$\log_b(3) = 0.57$$
    $$\log_b(5) = 0.84.$$

    a. Use the given values of $\log_b(2)$, $\log_b(3)$, and $\log_b(5)$ to evaluate the following logari
       i. $\log_b(6)$
       ii. $\log_b(8)$
       iii. $\log_b(10)$
       iv. $\log_b(600)$

    b. Use the change of base formula to convert $\log_b(10)$ to base 10, and solve for b. Give
       decimal places.

14. Use a logarithm with an appropriate base to solve the following exponential equations.

    a. $2^{3x} = 16$

    b. $2^{x+3} = 4^{3x}$

    c. $3^{4x-2} = 27^{x+2}$

    d. $4^{2x} = \left(\frac{1}{4}\right)^{3x}$

    e. $5^{0.2x+3} = 625$

15.

16.

17.

S.96

Lesson 13: Changing the Base

EUREKA MATH

©2015 Great Minds. eureka-math.org
ALG II-M3-SE-B2-1.3.0-08.2015

EUREKA MATH

Lesson 13: Changing the Base

S.98

©2015 Great Minds. eureka-math.org
ALG II-M3-SE-B2-1.3.0-08.2015

Lesson Summary

We have established a formula for changing the base of logarithms from b to a:

$$\log_b(x) = \frac{\log_a(x)}{\log_a(b)}.$$

In particular, the formula allows us to change logarithms base b to common or natural logarithms, which are the only two kinds of logarithms that most calculators compute:

$$\log_b(x) = \frac{\log(x)}{\log(b)} = \frac{\ln(x)}{\ln(b)}.$$

We have also established the following properties for base b logarithms. If x, y, a, and b are all positive real numbers with $a \neq 1$ and $b \neq 1$ and r is any real number, then:

1. $\log_b(1) = 0$
2. $\log_b(b) = 1$
3. $\log_b(b^r) = r$
4. $b^{\log_b(x)} = x$
5. $\log_b(x \cdot y) = \log_b(x) + \log_b(y)$
6. $\log_b(x^r) = r \cdot \log_b(x)$
7. $\log_b\left(\frac{1}{x}\right) = -\log_b(x)$
8. $\log_b\left(\frac{x}{y}\right) = \log_b(x) - \log_b(y)$

Problem Set

1. Evaluate each of the following logarithmic expressions, approximating to four decimal places if necessary. Use the LN or LOG key on your calculator rather than a table.

   a. $\log_8(16)$

   b. $\log_7(11)$

   c. $\log_3(2) + \log_2(3)$

2. Use logarithmic properties and the fact that $\ln(2) \approx 0.69$ and $\ln(3) \approx 1.10$ to approximate the value of each of the following logarithmic expressions. Do not use a calculator.

   a. $\ln(e^4)$

   b. $\ln(6)$

   c. $\ln(108)$

   d. $\ln\left(\frac{8}{3}\right)$

3. Compare the values of $\log_{\frac{1}{9}}(10)$ and $\log_9\left(\frac{1}{10}\right)$ without using a calculator.

4. Show that for any positive numbers a and b with $a \neq 1$ and $b \neq 1$, $\log_a(b) \cdot \log_b(a) = 1$.

5. Express x in terms of a, e, and y if $\ln(x) - \ln(y) = 2a$.

6. Rewrite each expression in an equivalent form that only contains one base 10 logarithm.

    a. $\log_2(800)$

    b. $\log_x\left(\frac{1}{10}\right)$, for positive real values of $x \neq 1$

    c. $\log_5(12500)$

    d. $\log_3(0.81)$

7. Write each number in terms of natural logarithms, and then use the properties of logarithms to show that it is a rational number.

    a. $\log_9\left(\sqrt{27}\right)$

    b. $\log_8(32)$

    c. $\log_4\left(\frac{1}{8}\right)$

8. Write each expression as an equivalent expression with a single logarithm. Assume x, y, and z are positive real numbers.

    a. $\ln(x) + 2\ln(y) - 3\ln(z)$

    b. $\frac{1}{2}\left(\ln(x + y) - \ln(z)\right)$

    c. $(x + y) + \ln(z)$

9. Rewrite each expression as sums and differences in terms of $\ln(x)$, $\ln(y)$, and $\ln(z)$.

    a. $\ln(xyz^3)$

    b. $\ln\left(\frac{e^3}{xyz}\right)$

    c. $\ln\left(\sqrt{\frac{x}{y}}\right)$

10. Use base-5 logarithms to rewrite each exponential equation as a logarithmic equation, and solve the resulting equation. Use the change of base formula to convert to a base-10 logarithm that can be evaluated on a calculator. Give each answer to 4 decimal places. If an equation has no solution, explain why.

   a. $5^{2x} = 20$

   b. $75 = 10 \cdot 5^{x-1}$

   c. $5^{2+x} - 5^x = 10$

   d. $5^{x^2} = 0.25$

11. In Lesson 6, you discovered that $\log(x \cdot 10^k) = k + \log(x)$ by looking at a table of logarithms. Use the properties of logarithms to justify this property for an arbitrary base $b > 0$ with $b \neq 1$. That is, show that $\log_b(x \cdot b^k) = k + \log_b(x)$.

12. Larissa argued that since $\log_2(2) = 1$ and $\log_2(4) = 2$, then it must be true that $\log_2(3) = 1.5$. Is she correct? Explain how you know.

13. Extension: Suppose that there is some positive number b so that

$$\log_b(2) = 0.36$$

$$\log_b(3) = 0.57$$

$$\log_b(5) = 0.84.$$

   a. Use the given values of $\log_b(2)$, $\log_b(3)$, and $\log_b(5)$ to evaluate the following logarithms:

      i. $\log_b(6)$

      ii. $\log_b(8)$

      iii. $\log_b(10)$

      iv. $\log_b(600)$

   b. Use the change of base formula to convert $\log_b(10)$ to base 10, and solve for b. Give your answer to four decimal places.

14. Use a logarithm with an appropriate base to solve the following exponential equations.

   a. $2^{3x} = 16$

   b. $2^{x+3} = 4^{3x}$

   c. $3^{4x-2} = 27^{x+2}$

   d. $4^{2x} = \left(\frac{1}{4}\right)^{3x}$

   e. $5^{0.2x+3} = 625$

15. Solve each exponential equation.

a. $3^{2x} = 81$ h. $2^x = 81$

b. $6^{3x} = 36^{x+1}$ i. $8 = 3^x$

c. $625 = 5^{3x}$ j. $6^{x+2} = 12$

d. $25^{4-x} = 5^{3x}$ k. $10^{x+4} = 27$

e. $32^{x-1} = \dfrac{1}{2}$ l. $2^{x+1} = 3^{1-x}$

                                          m. $3^{2x-3} = 2^{x+4}$

f. $\dfrac{4^{2x}}{2^{x-3}} = 1$ n. $e^{2x} = 5$

                                          o. $e^{x-1} = 6$

g. $\dfrac{1}{8^{2x-4}} = 1$

16. In Problem 9(e) of Lesson 12, you solved the equation $3^x = 7^{-3x+2}$ using the logarithm base 10.

a. Solve $3^x = 7^{-3x+2}$ using the logarithm base 3.

b. Apply the change of base formula to show that your answer to part (a) agrees with your answer to Problem 9(e) of Lesson 12.

c. Solve $3^x = 7^{-3x+2}$ using the logarithm base 7.

d. Apply the change of base formula to show that your answer to part (c) also agrees with your answer to Problem 9(e) of Lesson 12.

17. Pearl solved the equation $2^x = 10$ as follows:

$$\log(2^x) = \log(10)$$

$$x \log(2) = 1$$

$$x = \frac{1}{\log(2)}.$$

Jess solved the equation $2^x = 10$ as follows:

$$\log_2(2^x) = \log_2(10)$$

$$x \log_2(2) = \log_2(10)$$

$$x = \log_2(10).$$

Is Pearl correct? Is Jess correct? Explain how you know.

Lesson 14: Solving Logarithmic Equations

Classwork

Opening Exercise

Convert the following logarithmic equations to equivalent exponential equations.

a. $\log(10,000) = 4$

b. $\log(\sqrt{10}) = \frac{1}{2}$

c. $\log_2(256) = 8$

d. $\log_4(256) = 4$

e. $\ln(1) = 0$

f. $\log(x + 2) = 3$

Examples

Write each of the following equations as an equivalent exponential equation, and solve for x.

1. $\log(3x + 7) = 0$

2. $\log_2(x + 5) = 4$

3. $\log(x + 2) + \log(x + 5) = 1$

Exercises

1. Drew said that the equation $\log_2[(x+1)^4] = 8$ cannot be solved because he expanded $(x+1)^4 = x^4 + 4x^3 + 6x^2 + 4x + 1$ and realized that he cannot solve the equation $x^4 + 4x^3 + 6x^2 + 4x + 1 = 2^8$. Is he correct? Explain how you know.

Solve the equations in Exercises 2–4 for x.

2. $\ln((4x)^5) = 15$

3. $\log((2x+5)^2) = 4$

4. $\log_2((5x + 7)^{19}) = 57$

Solve the logarithmic equations in Exercises 5–9, and identify any extraneous solutions.

5. $\log(x^2 + 7x + 12) - \log(x + 4) = 0$

6. $\log_2(3x) + \log_2(4) = 4$

Lesson 14:   Solving Logarithmic Equations

©2015 Great Minds. eureka-math.org
ALG II-M3-SE-B2-1.3.0-08.2015

7. $2\ln(x + 2) - \ln(-x) = 0$

8. $\log(x) = 2 - \log(x)$

9. $\ln(x + 2) = \ln(12) - \ln(x + 3)$

Problem Set

1. Solve the following logarithmic equations.

   a. $\log(x) = \dfrac{5}{2}$

   b. $5\log(x + 4) = 10$

   c. $\log_2(1 - x) = 4$

   d. $\log_2(49x^2) = 4$

   e. $\log_2(9x^2 + 30x + 25) = 8$

2. Solve the following logarithmic equations.

   a. $\ln(x^6) = 36$

   b. $\log[(2x^2 + 45x - 25)^5] = 10$

   c. $\log[(x^2 + 2x - 3)^4] = 0$

3. Solve the following logarithmic equations.

   a. $\log(x) + \log(x - 1) = \log(3x + 12)$

   b. $\ln(32x^2) - 3\ln(2) = 3$

   c. $\log(x) + \log(-x) = 0$

   d. $\log(x + 3) + \log(x + 5) = 2$

   e. $\log(10x + 5) - 3 = \log(x - 5)$

   f. $\log_2(x) + \log_2(2x) + \log_2(3x) + \log_2(36) = 6$

4. Solve the following equations.

   a. $\log_2(x) = 4$

   b. $\log_6(x) = 1$

   c. $\log_3(x) = -4$

   d. $\log_{\sqrt{2}}(x) = 4$

   e. $\log_{\sqrt{5}}(x) = 3$

   f. $\log_3(x^2) = 4$

   g. $\log_2(x^{-3}) = 12$

   h. $\log_3(8x + 9) = 4$

   i. $2 = \log_4(3x - 2)$

   j. $\log_5(3 - 2x) = 0$

   k. $\ln(2x) = 3$

   l. $\log_3(x^2 - 3x + 5) = 2$

   m. $\log((x^2 + 4)^5) = 10$

   n. $\log(x) + \log(x + 21) = 2$

   o. $\log_4(x - 2) + \log_4(2x) = 2$

   p. $\log(x) - \log(x + 3) = -1$

   q. $\log_4(x + 3) - \log_4(x - 5) = 2$

   r. $\log(x) + 1 = \log(x + 9)$

   s. $\log_3(x^2 - 9) - \log_3(x + 3) = 1$

   t. $1 - \log_8(x - 3) = \log_8(2x)$

   u. $\log_2(x^2 - 16) - \log_2(x - 4) = 1$

   v. $\log\left(\sqrt{(x + 3)^3}\right) = \dfrac{3}{2}$

   w. $\ln(4x^2 - 1) = 0$

   x. $\ln(x + 1) - \ln(2) = 1$

EUREKA
MATH™

Lesson 15: Why Were Logarithms Developed?

Classwork

Exercises

1. Solve the following equations. Remember to check for extraneous solutions because logarithms are only defined for positive real numbers.

    a. $\log(x^2) = \log(49)$

    b. $\log(x + 1) + \log(x - 2) = \log(7x - 17)$

    c. $\log(x^2 + 1) = \log(x(x - 2))$

d. $\log(x + 4) + \log(x - 1) = \log(3x)$

e. $\log(x^2 - x) - \log(x - 2) = \log(x - 3)$

f. $\log(x) + \log(x - 1) + \log(x + 1) = 3\log(x)$

EUREKA
MATH™

g. $\log(x - 4) = -\log(x - 2)$

2.   How do you know if you need to use the definition of logarithm to solve an equation involving logarithms as we did in Lesson 15 or if you can use the methods of this lesson?

Lesson Summary

A table of base-10 logarithms can be used to simplify multiplication of multi-digit numbers:

1. To compute $A \cdot B$ for positive real numbers A and B, look up $\log(A)$ and $\log(B)$ in the logarithm table.

2. Add $\log(A)$ and $\log(B)$. The sum can be written as $k + d$, where k is an integer and $0 \leq d < 1$ is the decimal part.

3. Look back at the table, and find the entry closest to the decimal part, d.

4. The product of that entry and 10^k is an approximation to $A \cdot B$.

A similar process simplifies division of multi-digit numbers:

1. To compute $\dfrac{A}{B}$ for positive real numbers A and B, look up $\log(A)$ and $\log(B)$ in the logarithm table.

2. Calculate $\log(A) - \log(B)$. The difference can be written as $k + d$, where k is an integer and $0 \leq d < 1$ is the decimal part.

3. Look back at the table to find the entry closest to the decimal part, d.

4. The product of that entry and 10^k is an approximation to $\dfrac{A}{B}$.

For any positive values X and Y, if $\log_b(X) = \log_b(Y)$, we can conclude that $X = Y$. This property is the essence of how a logarithm table works, and it allows us to solve equations with logarithmic expressions on both sides of the equation.

Problem Set

1. Use the table of logarithms to approximate solutions to the following logarithmic equations.

    a. $\log(x) = 0.5044$

    b. $\log(x) = -0.5044$ (Hint: Begin by writing -0.5044 as $[(-0.5044) + 1] - 1$.)

    c. $\log(x) = 35.5044$

    d. $\log(x) = 4.9201$

2. Use logarithms and the logarithm table to evaluate each expression.

    a. $\sqrt{2.33}$

    b. $13500 \cdot 3600$

    c. $\dfrac{7.2 \times 10^9}{1.3 \times 10^5}$

3. Solve for x: $\log(3) + 2\log(x) = \log(27)$.

©2015 Great Minds. eureka-math.org
ALG II-M3-SE-B2-1.3.0-08.2015

4. Solve for x: $\log(3x) + \log(x + 4) = \log(15)$.

5. Solve for x.
   a. $\log(x) = \log(y + z) + \log(y - z)$
   b. $\log(x) = \big(\log(y) + \log(z)\big) + \big(\log(y) - \log(z)\big)$

6. If x and y are positive real numbers, and $\log(y) = 1 + \log(x)$, express y in terms of x.

7. If x, y, and z are positive real numbers, and $\log(x) - \log(y) = \log(y) - \log(z)$, express y in terms of x and z.

8. If x and y are positive real numbers, and $\log(x) = y\big(\log(y + 1) - \log(y)\big)$, express x in terms of y.

9. If x and y are positive real numbers, and $\log(y) = 3 + 2\log(x)$, express y in terms of x.

10. If x, y, and z are positive real numbers, and $\log(z) = \log(y) + 2\log(x) - 1$, express z in terms of x and y.

11. Solve the following equations.
    a. $\ln(10) - \ln(7 - x) = \ln(x)$
    b. $\ln(x + 2) + \ln(x - 2) = \ln(9x - 24)$
    c. $\ln(x + 2) + \ln(x - 2) = \ln(-2x - 1)$

12. Suppose the formula $P = P_0(1 + r)^t$ gives the population of a city P growing at an annual percent rate r, where P_0 is the population t years ago.
    a. Find the time t it takes this population to double.
    b. Use the structure of the expression to explain why populations with lower growth rates take a longer time to double.
    c. Use the structure of the expression to explain why the only way to double the population in one year is if there is a 100 percent growth rate.

13. If $x > 0$, $a + b > 0$, $a > b$, and $\log(x) = \log(a + b) + \log(a - b)$, find x in terms of a and b.

14. Jenn claims that because $\log(1) + \log(2) + \log(3) = \log(6)$, then $\log(2) + \log(3) + \log(4) = \log(9)$.
    a. Is she correct? Explain how you know.
    b. If $\log(a) + \log(b) + \log(c) = \log(a + b + c)$, express c in terms of a and b. Explain how this result relates to your answer to part (a).
    c. Find other values of a, b, and c so that $\log(a) + \log(b) + \log(c) = \log(a + b + c)$.

15. In Problem 7 of the Lesson 12 Problem Set, you showed that for $x \geq 1$, $\log(x + \sqrt{x^2 - 1}) + \log(x - \sqrt{x^2 - 1}) = 0$. It follows that $\log(x + \sqrt{x^2 - 1}) = -\log(x - \sqrt{x^2 - 1})$. What does this tell us about the relationship between the expressions $x + \sqrt{x^2 - 1}$ and $x - \sqrt{x^2 - 1}$?

16. Use the change of base formula to solve the following equations.

a. $\log(x) = \log_{100}(x^2 - 2x + 6)$

b. $\log(x - 2) = \log_{100}(14 - x)$

c. $\log_2(x + 1) = \log_4(x^2 + 3x + 4)$

d. $\log_2(x - 1) = \log_8(x^3 - 2x^2 - 2x + 5)$

17. Solve the following equation: $\log(9x) = \dfrac{2\ln(3) + \ln(x)}{\ln(10)}$.

EUREKA
MATH™

Common Logarithm Table

N	0	1	2	3	4	5	6	7	8	9
1.0	0.0000	0.0043	0.0086	0.0128	0.0170	0.0212	0.0253	0.0294	0.0334	0.0374
1.1	0.0414	0.0453	0.0492	0.0531	0.0569	0.0607	0.0645	0.0682	0.0719	0.0755
1.2	0.0793	0.0828	0.0864	0.0899	0.0934	0.0969	0.1004	0.1038	0.1072	0.1106
1.3	0.1142	0.1173	0.1206	0.1239	0.1271	0.1303	0.1335	0.1367	0.1399	0.1430
1.4	0.1465	0.1492	0.1523	0.1553	0.1584	0.1614	0.1644	0.1673	0.1703	0.1732
1.5	0.1765	0.1790	0.1818	0.1847	0.1875	0.1903	0.1931	0.1959	0.1987	0.2014
1.6	0.2046	0.2068	0.2095	0.2122	0.2148	0.2175	0.2201	0.2227	0.2253	0.2279
1.7	0.2310	0.2330	0.2355	0.2380	0.2405	0.2430	0.2455	0.2480	0.2504	0.2529
1.8	0.2558	0.2577	0.2601	0.2625	0.2648	0.2672	0.2695	0.2718	0.2742	0.2765
1.9	0.2793	0.2810	0.2833	0.2856	0.2878	0.2900	0.2923	0.2945	0.2967	0.2989
2.0	0.3016	0.3032	0.3054	0.3075	0.3096	0.3118	0.3139	0.3160	0.3181	0.3201
2.1	0.3228	0.3243	0.3263	0.3284	0.3304	0.3324	0.3345	0.3365	0.3385	0.3404
2.2	0.3431	0.3444	0.3464	0.3483	0.3502	0.3522	0.3541	0.3560	0.3579	0.3598
2.3	0.3624	0.3636	0.3655	0.3674	0.3692	0.3711	0.3729	0.3747	0.3766	0.3784
2.4	0.3809	0.3820	0.3838	0.3856	0.3874	0.3892	0.3909	0.3927	0.3945	0.3962
2.5	0.3986	0.3997	0.4014	0.4031	0.4048	0.4065	0.4082	0.4099	0.4116	0.4133
2.6	0.4156	0.4166	0.4183	0.4200	0.4216	0.4232	0.4249	0.4265	0.4281	0.4298
2.7	0.4320	0.4330	0.4346	0.4362	0.4378	0.4393	0.4409	0.4425	0.4440	0.4456
2.8	0.4478	0.4487	0.4502	0.4518	0.4533	0.4548	0.4564	0.4579	0.4594	0.4609
2.9	0.4631	0.4639	0.4654	0.4669	0.4683	0.4698	0.4713	0.4728	0.4742	0.4757
3.0	0.4778	0.4786	0.4800	0.4814	0.4829	0.4843	0.4857	0.4871	0.4886	0.4900
3.1	0.4920	0.4928	0.4942	0.4955	0.4969	0.4983	0.4997	0.5011	0.5024	0.5038
3.2	0.5058	0.5065	0.5079	0.5092	0.5105	0.5119	0.5132	0.5145	0.5159	0.5172
3.3	0.5192	0.5198	0.5211	0.5224	0.5237	0.5250	0.5263	0.5276	0.5289	0.5302
3.4	0.5321	0.5328	0.5340	0.5353	0.5366	0.5378	0.5391	0.5403	0.5416	0.5428
3.5	0.5447	0.5453	0.5465	0.5478	0.5490	0.5502	0.5514	0.5527	0.5539	0.5551
3.6	0.5570	0.5575	0.5587	0.5599	0.5611	0.5623	0.5635	0.5647	0.5658	0.5670
3.7	0.5689	0.5694	0.5705	0.5717	0.5729	0.5740	0.5752	0.5763	0.5775	0.5786
3.8	0.5804	0.5809	0.5821	0.5832	0.5843	0.5855	0.5866	0.5877	0.5888	0.5899
3.9	0.5917	0.5922	0.5933	0.5944	0.5955	0.5966	0.5977	0.5988	0.5999	0.6010
4.0	0.6027	0.6031	0.6042	0.6053	0.6064	0.6075	0.6085	0.6096	0.6107	0.6117
4.1	0.6134	0.6138	0.6149	0.6160	0.6170	0.6180	0.6191	0.6201	0.6212	0.6222
4.2	0.6239	0.6243	0.6253	0.6263	0.6274	0.6284	0.6294	0.6304	0.6314	0.6325
4.3	0.6341	0.6345	0.6355	0.6365	0.6375	0.6385	0.6395	0.6405	0.6415	0.6425
4.4	0.6441	0.6444	0.6454	0.6464	0.6474	0.6484	0.6493	0.6503	0.6513	0.6522
4.5	0.6538	0.6542	0.6551	0.6561	0.6571	0.6580	0.6590	0.6599	0.6609	0.6618
4.6	0.6634	0.6637	0.6646	0.6656	0.6665	0.6675	0.6684	0.6693	0.6702	0.6712
4.7	0.6727	0.6730	0.6739	0.6749	0.6758	0.6767	0.6776	0.6785	0.6794	0.6803
4.8	0.6818	0.6821	0.6830	0.6839	0.6848	0.6857	0.6866	0.6875	0.6884	0.6893
4.9	0.6908	0.6911	0.6920	0.6928	0.6937	0.6946	0.6955	0.6964	0.6972	0.6981
5.0	0.6996	0.6998	0.7007	0.7016	0.7024	0.7033	0.7042	0.7050	0.7059	0.7067
5.1	0.7082	0.7084	0.7093	0.7101	0.7110	0.7118	0.7126	0.7135	0.7143	0.7152
5.2	0.7166	0.7168	0.7177	0.7185	0.7193	0.7202	0.7210	0.7218	0.7226	0.7235
5.3	0.7249	0.7251	0.7259	0.7267	0.7275	0.7284	0.7292	0.7300	0.7308	0.7316
5.4	0.7330	0.7332	0.7340	0.7348	0.7356	0.7364	0.7372	0.7380	0.7388	0.7396

N	0	1	2	3	4	5	6	7	8	9
5.5	0.7404	0.7412	0.7419	0.7427	0.7435	0.7443	0.7451	0.7459	0.7466	0.7474
5.6	0.7482	0.7490	0.7497	0.7505	0.7513	0.7520	0.7528	0.7536	0.7543	0.7551
5.7	0.7559	0.7566	0.7574	0.7582	0.7589	0.7597	0.7604	0.7612	0.7619	0.7627
5.8	0.7634	0.7642	0.7649	0.7657	0.7664	0.7672	0.7679	0.7686	0.7694	0.7701
5.9	0.7709	0.7716	0.7723	0.7731	0.7738	0.7745	0.7752	0.7760	0.7767	0.7774
6.0	0.7782	0.7789	0.7796	0.7803	0.7810	0.7818	0.7825	0.7832	0.7839	0.7846
6.1	0.7853	0.7860	0.7868	0.7875	0.7882	0.7889	0.7896	0.7903	0.7910	0.7917
6.2	0.7924	0.7931	0.7938	0.7945	0.7952	0.7959	0.7966	0.7973	0.7980	0.7987
6.3	0.7993	0.8000	0.8007	0.8014	0.8021	0.8028	0.8035	0.8041	0.8048	0.8055
6.4	0.8062	0.8069	0.8075	0.8082	0.8089	0.8096	0.8102	0.8109	0.8116	0.8122
6.5	0.8129	0.8136	0.8142	0.8149	0.8156	0.8162	0.8169	0.8176	0.8182	0.8189
6.6	0.8195	0.8202	0.8209	0.8215	0.8222	0.8228	0.8235	0.8241	0.8248	0.8254
6.7	0.8261	0.8267	0.8274	0.8280	0.8287	0.8293	0.8299	0.8306	0.8312	0.8319
6.8	0.8325	0.8331	0.8338	0.8344	0.8351	0.8357	0.8363	0.8370	0.8376	0.8382
6.9	0.8388	0.8395	0.8401	0.8407	0.8414	0.8420	0.8426	0.8432	0.8439	0.8445
7.0	0.8451	0.8457	0.8463	0.8470	0.8476	0.8482	0.8488	0.8494	0.8500	0.8506
7.1	0.8513	0.8519	0.8525	0.8531	0.8537	0.8543	0.8549	0.8555	0.8561	0.8567
7.2	0.8573	0.8579	0.8585	0.8591	0.8597	0.8603	0.8609	0.8615	0.8621	0.8627
7.3	0.8633	0.8639	0.8645	0.8651	0.8657	0.8663	0.8669	0.8675	0.8681	0.8686
7.4	0.8692	0.8698	0.8704	0.8710	0.8716	0.8722	0.8727	0.8733	0.8739	0.8745
7.5	0.8751	0.8756	0.8762	0.8768	0.8774	0.8779	0.8785	0.8791	0.8797	0.8802
7.6	0.8808	0.8814	0.8820	0.8825	0.8831	0.8837	0.8842	0.8848	0.8854	0.8859
7.7	0.8865	0.8871	0.8876	0.8882	0.8887	0.8893	0.8899	0.8904	0.8910	0.8915
7.8	0.8921	0.8927	0.8932	0.8938	0.8943	0.8949	0.8954	0.8960	0.8965	0.8971
7.9	0.8976	0.8982	0.8987	0.8993	0.8998	0.9004	0.9009	0.9015	0.9020	0.9025
8.0	0.9031	0.9036	0.9042	0.9047	0.9053	0.9058	0.9063	0.9069	0.9074	0.9079
8.1	0.9085	0.9090	0.9096	0.9101	0.9106	0.9112	0.9117	0.9122	0.9128	0.9133
8.2	0.9138	0.9143	0.9149	0.9154	0.9159	0.9165	0.9170	0.9175	0.9180	0.9186
8.3	0.9191	0.9196	0.9201	0.9206	0.9212	0.9217	0.9222	0.9227	0.9232	0.9238
8.4	0.9243	0.9248	0.9253	0.9258	0.9263	0.9269	0.9274	0.9279	0.9284	0.9289
8.5	0.9294	0.9299	0.9304	0.9309	0.9315	0.9320	0.9325	0.9330	0.9335	0.9340
8.6	0.9345	0.9350	0.9355	0.9360	0.9365	0.9370	0.9375	0.9380	0.9385	0.9390
8.7	0.9395	0.9400	0.9405	0.9410	0.9415	0.9420	0.9425	0.9430	0.9435	0.9440
8.8	0.9445	0.9450	0.9455	0.9460	0.9465	0.9469	0.9474	0.9479	0.9484	0.9489
8.9	0.9494	0.9499	0.9504	0.9509	0.9513	0.9518	0.9523	0.9528	0.9533	0.9538
9.0	0.9542	0.9547	0.9552	0.9557	0.9562	0.9566	0.9571	0.9576	0.9581	0.9586
9.1	0.9590	0.9595	0.9600	0.9605	0.9609	0.9614	0.9619	0.9624	0.9628	0.9633
9.2	0.9638	0.9643	0.9647	0.9652	0.9657	0.9661	0.9666	0.9671	0.9675	0.9680
9.3	0.9685	0.9689	0.9694	0.9699	0.9703	0.9708	0.9713	0.9717	0.9722	0.9727
9.4	0.9731	0.9736	0.9741	0.9745	0.9750	0.9754	0.9759	0.9763	0.9768	0.9773
9.5	0.9777	0.9782	0.9786	0.9791	0.9795	0.9800	0.9805	0.9809	0.9814	0.9818
9.6	0.9823	0.9827	0.9832	0.9836	0.9841	0.9845	0.9850	0.9854	0.9859	0.9863
9.7	0.9868	0.9872	0.9877	0.9881	0.9886	0.9890	0.9894	0.9899	0.9903	0.9908
9.8	0.9912	0.9917	0.9921	0.9926	0.9930	0.9934	0.9939	0.9943	0.9948	0.9952
9.9	0.9956	0.9961	0.9965	0.9969	0.9974	0.9978	0.9983	0.9987	0.9991	0.9996

Lesson 16: Rational and Irrational Numbers

Classwork

Opening Exercise

a. Explain how to use a number line to add the fractions $\frac{7}{5} + \frac{9}{4}$.

b. Convert $\frac{7}{5}$ and $\frac{9}{4}$ to decimals, and explain the process for adding them together.

Exercises

1. According to the calculator, $\log(4) = 0.6020599913\ldots$ and $\log(25) = 1.3979400087\ldots$ Find an approximation of $\log(4) + \log(25)$ to one decimal place, that is, to an accuracy of 10^{-1}.

2. Find the value of $\log(4) + \log(25)$ to an accuracy of 10^{-2}.

3. Find the value of $\log(4) + \log(25)$ to an accuracy of 10^{-8}.

4. Make a conjecture: Is $\log(4) + \log(25)$ a rational or an irrational number?

5. Why is your conjecture in Exercise 4 true?

Remember that the calculator gives the following values: $\log(4) = 0.6020599913\ldots$ and $\log(25) = 1.3979400087\ldots$

6. Find the value of $\log(4) \cdot \log(25)$ to three decimal places.

7. Find the value of $\log(4) \cdot \log(25)$ to five decimal places.

8. Does your conjecture from the above discussion appear to be true?

Lesson Summary

- Irrational numbers occur naturally and frequently.
- The nth roots of most integers and rational numbers are irrational.
- Logarithms of most positive integers or positive rational numbers are irrational.
- We can locate an irrational number on the number line by trapping it between lower and upper estimates. The infinite process of squeezing the irrational number in smaller and smaller intervals locates exactly where the irrational number is on the number line.
- We can perform arithmetic operations such as addition and multiplication with irrational numbers using lower and upper approximations and squeezing the result of the operation in smaller and smaller intervals between two rational approximations to the result.

Problem Set

1. Given that $\sqrt{5} \approx 2.2360679775$ and $\pi \approx 3.1415926535$, find the sum $\sqrt{5} + \pi$ to an accuracy of 10^{-8} without using a calculator.

2. Put the following numbers in order from least to greatest.

$$\sqrt{2}, \pi, 0, e, \frac{22}{7}, \frac{\pi^2}{3}, 3.14, \sqrt{10}$$

3. Find a rational number between the specified two numbers.

   a. $\frac{4}{13}$ and $\frac{5}{13}$

   b. $\frac{3}{8}$ and $\frac{5}{9}$

   c. 1.7299999 and 1.73

   d. $\frac{\sqrt{2}}{7}$ and $\frac{\sqrt{2}}{9}$

   e. π and $\sqrt{10}$

4. Knowing that $\sqrt{2}$ is irrational, find an irrational number between $\frac{1}{2}$ and $\frac{5}{9}$.

5. Give an example of an irrational number between e and π.

6. Given that $\sqrt{2}$ is irrational, which of the following numbers are irrational?

$$\frac{\sqrt{2}}{2}, 2 + \sqrt{2}, \frac{\sqrt{2}}{2\sqrt{2}}, \frac{2}{\sqrt{2}}, \left(\sqrt{2}\right)^2$$

7. Given that π is irrational, which of the following numbers are irrational?

$$\frac{\pi}{2}, \frac{\pi}{2\pi}, \sqrt{\pi}, \pi^2$$

8. Which of the following numbers are irrational?

$$1, 0, \sqrt{5}, \sqrt[3]{64}, e, \pi, \frac{\sqrt{2}}{2}, \frac{\sqrt{8}}{\sqrt{2}}, \cos\left(\frac{\pi}{3}\right), \sin\left(\frac{\pi}{3}\right)$$

9. Find two irrational numbers x and y so that their average is rational.

10. Suppose that $\frac{2}{3}x$ is an irrational number. Explain how you know that x must be an irrational number. (Hint: What would happen if there were integers a and b so that $x = \frac{a}{b}$?)

11. If r and s are rational numbers, prove that $r + s$ and $r - s$ are also rational numbers.

12. If r is a rational number and x is an irrational number, determine whether the following numbers are always rational, sometimes rational, or never rational. Explain how you know.

   a. $r + x$

   b. $r - x$

   c. rx

   d. x^r

13. If x and y are irrational numbers, determine whether the following numbers are always rational, sometimes rational, or never rational. Explain how you know.

   a. $x + y$

   b. $x - y$

   c. xy

   d. $\dfrac{x}{y}$

This page intentionally left blank

Lesson 17: Graphing the Logarithm Function

Classwork

Opening Exercise

Graph the points in the table for your assigned function $f(x) = \log(x)$, $g(x) = \log_2(x)$, or $h(x) = \log_5(x)$ for $0 < x \le 16$. Then, sketch a smooth curve through those points and answer the questions that follow.

10-team		2-team		5-team	
$f(x) = \log(x)$		$g(x) = \log_2(x)$		$h(x) = \log_5(x)$	
x	$f(x)$	x	$g(x)$	x	$h(x)$
0.0625	−1.20	0.0625	−4	0.0625	−1.72
0.125	−0.90	0.125	−3	0.125	−1.29
0.25	−0.60	0.25	−2	0.25	−0.86
0.5	−0.30	0.5	−1	0.5	−0.43
1	0	1	0	1	0
2	0.30	2	1	2	0.43
4	0.60	4	2	4	0.86
8	0.90	8	3	8	1.29
16	1.20	16	4	16	1.72

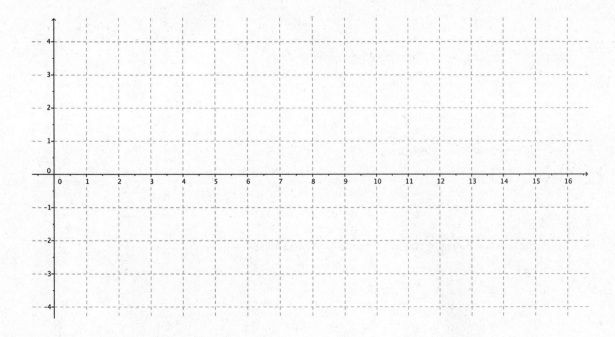

a. What does the graph indicate about the domain of your function?

b. Describe the x-intercepts of the graph.

c. Describe the y-intercepts of the graph.

d. Find the coordinates of the point on the graph with y-value 1.

e. Describe the behavior of the function as $x \to 0$.

f. Describe the end behavior of the function as $x \to \infty$.

g. Describe the range of your function.

h. Does this function have any relative maxima or minima? Explain how you know.

EUREKA
MATH™

Exercises

1. Graph the points in the table for your assigned function $r(x) = \log_{\frac{1}{10}}(x)$, $s(x) = \log_{\frac{1}{2}}(x)$, or $t(x) = \log_{\frac{1}{5}}(x)$ for $0 < x \le 16$. Then sketch a smooth curve through those points, and answer the questions that follow.

10-team		2-team		e-team	
$r(x) = \log_{\frac{1}{10}}(x)$		$s(x) = \log_{\frac{1}{2}}(x)$		$t(x) = \log_{\frac{1}{5}}(x)$	
x	$r(x)$	x	$s(x)$	x	$t(x)$
0.0625	1.20	0.0625	4	0.0625	1.72
0.125	0.90	0.125	3	0.125	1.29
0.25	0.60	0.25	2	0.25	0.86
0.5	0.30	0.5	1	0.5	0.43
1	0	1	0	1	0
2	−0.30	2	−1	2	−0.43
4	−0.60	4	−2	4	−0.86
8	−0.90	8	−3	8	−1.29
16	−1.20	16	−4	16	−1.72

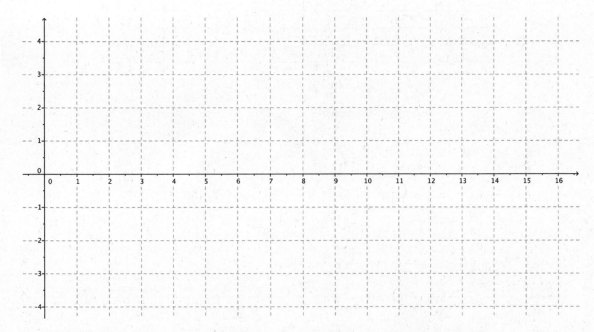

a. What is the relationship between your graph in the Opening Exercise and your graph from this exercise?

b. Why does this happen? Use the change of base formula to justify what you have observed in part (a).

2. In general, what is the relationship between the graph of a function $y = f(x)$ and the graph of $y = f(kx)$ for a constant k?

3. Graph the points in the table for your assigned function $u(x) = \log(10x)$, $v(x) = \log_2(2x)$, or $w(x) = \log_5(5x)$ for $0 < x \le 16$. Then sketch a smooth curve through those points, and answer the questions that follow.

10-team		2-team		5-team	
$u(x) = \log(10x)$		$v(x) = \log_2(2x)$		$w(x) = \log_5(5x)$	
x	$u(x)$	x	$v(x)$	x	$w(x)$
0.0625	−0.20	0.0625	−3	0.0625	−0.72
0.125	0.10	0.125	−2	0.125	−0.29
0.25	0.40	0.25	−1	0.25	0.14
0.5	0.70	0.5	0	0.5	0.57
1	1	1	1	1	1
2	1.30	2	2	2	1.43
4	1.60	4	3	4	1.86
8	1.90	8	4	8	2.29
16	2.20	16	5	16	2.72

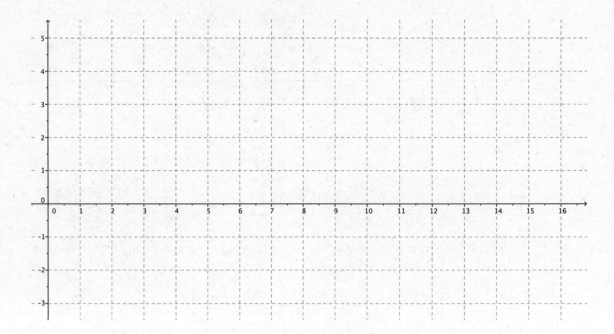

a. Describe a transformation that takes the graph of your team's function in this exercise to the graph of your team's function in the Opening Exercise.

b. Do your answers to Exercise 2 and part (a) agree? If not, use properties of logarithms to justify your observations in part (a).

Lesson Summary

The function $f(x) = \log_b(x)$ is defined for irrational and rational numbers. Its domain is all positive real numbers. Its range is all real numbers.

The function $f(x) = \log_b(x)$ goes to negative infinity as x goes to zero. It goes to positive infinity as x goes to positive infinity.

The larger the base b, the more slowly the function $f(x) = \log_b(x)$ increases.

By the change of base formula, $\log_{\frac{1}{b}}(x) = -\log_b(x)$.

Problem Set

1. The function $Q(x) = \log_b(x)$ has function values in the table at right.

   a. Use the values in the table to sketch the graph of $y = Q(x)$.

   b. What is the value of b in $Q(x) = \log_b(x)$? Explain how you know.

   c. Identify the key features in the graph of $y = Q(x)$.

x	$Q(x)$
0.1	1.66
0.3	0.87
0.5	0.50
1.00	0.00
2.00	−0.50
4.00	−1.00
6.00	−1.29
10.00	−1.66
12.00	−1.79

2. Consider the logarithmic functions $f(x) = \log_b(x)$, $g(x) = \log_5(x)$, where b is a positive real number, and $b \neq 1$. The graph of f is given at right.

   a. Is $b > 5$, or is $b < 5$? Explain how you know.

   b. Compare the domain and range of functions f and g.

   c. Compare the x-intercepts and y-intercepts of f and g.

   d. Compare the end behavior of f and g.

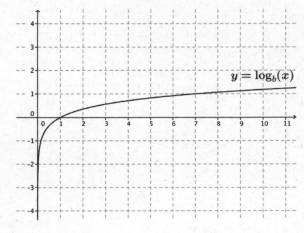

$y = \log_b(x)$

3. Consider the logarithmic functions $f(x) = \log_b(x)$, $g(x) = \log_{\frac{1}{2}}(x)$, where b is a positive real number and $b \neq 1$.

   A table of approximate values of f is given below.

   a. Is $b > \frac{1}{2}$, or is $b < \frac{1}{2}$? Explain how you know.

   b. Compare the domain and range of functions f and g.

   c. Compare the x-intercepts and y-intercepts of f and g.

   d. Compare the end behavior of f and g.

x	$f(x)$
$\frac{1}{4}$	0.86
$\frac{1}{2}$	0.43
1	0
2	−0.43
4	−0.86

4. On the same set of axes, sketch the functions $f(x) = \log_2(x)$ and $g(x) = \log_2(x^3)$.

   a. Describe a transformation that takes the graph of f to the graph of g.

   b. Use properties of logarithms to justify your observations in part (a).

5. On the same set of axes, sketch the functions $f(x) = \log_2(x)$ and $g(x) = \log_2\left(\frac{x}{4}\right)$.

   a. Describe a transformation that takes the graph of f to the graph of g.

   b. Use properties of logarithms to justify your observations in part (a).

6. On the same set of axes, sketch the functions $f(x) = \log_{\frac{1}{2}}(x)$ and $g(x) = \log_2\left(\frac{1}{x}\right)$.

   a. Describe a transformation that takes the graph of f to the graph of g.

   b. Use properties of logarithms to justify your observations in part (a).

7. The figure below shows graphs of the functions $f(x) = \log_3(x)$, $g(x) = \log_5(x)$, and $h(x) = \log_{11}(x)$.

   a. Identify which graph corresponds to which function. Explain how you know.

   b. Sketch the graph of $k(x) = \log_7(x)$ on the same axes.

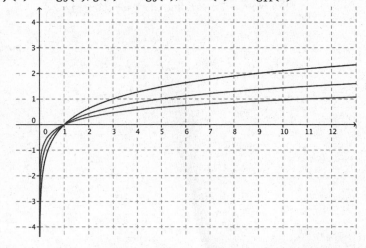

8. The figure below shows graphs of the functions $f(x) = \log_{\frac{1}{3}}(x)$, $g(x) = \log_{\frac{1}{5}}(x)$, and $h(x) = \log_{\frac{1}{11}}(x)$.

a. Identify which graph corresponds to which function. Explain how you know.

b. Sketch the graph of $k(x) = \log_{\frac{1}{7}}(x)$ on the same axes.

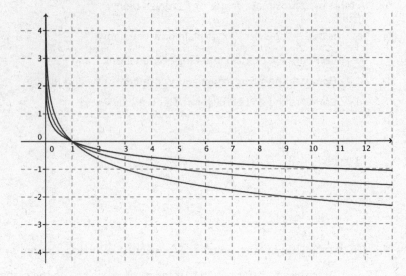

9. For each function f, find a formula for the function h in terms of x. Part (a) has been done for you.

a. If $f(x) = x^2 + x$, find $h(x) = f(x + 1)$.

b. If $f(x) = \sqrt{x^2 + \frac{1}{4}}$, find $h(x) = f\left(\frac{1}{2}x\right)$.

c. If $f(x) = \log(x)$, find $h(x) = f(\sqrt[3]{10x})$ when $x > 0$.

d. If $f(x) = 3^x$, find $h(x) = f(\log_3(x^2 + 3))$.

e. If $f(x) = x^3$, find $h(x) = f\left(\frac{1}{x^3}\right)$ when $x \neq 0$.

f. If $f(x) = x^3$, find $h(x) = f(\sqrt[3]{x})$.

g. If $f(x) = \sin(x)$, find $h(x) = f\left(x + \frac{\pi}{2}\right)$.

h. If $f(x) = x^2 + 2x + 2$, find $h(x) = f(\cos(x))$.

EUREKA
MATH™

10. For each of the functions f and g below, write an expression for (i) $f(g(x))$, (ii) $g(f(x))$, and (iii) $f(f(x))$ in terms of x. Part (a) has been done for you.

   a. $f(x) = x^2$, $g(x) = x + 1$

   i. $f(g(x)) = f(x + 1)$
       $$= (x + 1)^2$$

   ii. $g(f(x)) = g(x^2)$
        $$= x^2 + 1$$

   iii. $f(f(x)) = f(x^2)$
         $$= (x^2)^2$$
         $$= x^4$$

   b. $f(x) = \frac{1}{4}x - 8$, $g(x) = 4x + 1$

   c. $f(x) = \sqrt[3]{x + 1}$, $g(x) = x^3 - 1$

   d. $f(x) = x^3$, $g(x) = \frac{1}{x}$

   e. $f(x) = |x|$, $g(x) = x^2$

Extension:

11. Consider the functions $f(x) = \log_2(x)$ and $(x) = \sqrt{x - 1}$.

   a. Use a calculator or other graphing utility to produce graphs of $f(x) = \log_2(x)$ and $g(x) = \sqrt{x - 1}$ for $x \le 17$.

   b. Compare the graph of the function $f(x) = \log_2(x)$ with the graph of the function $g(x) = \sqrt{x - 1}$. Describe the similarities and differences between the graphs.

   c. Is it always the case that $\log_2(x) > \sqrt{x - 1}$ for $x > 2$?

12. Consider the functions $f(x) = \log_2(x)$ and $(x) = \sqrt[3]{x - 1}$.

   a. Use a calculator or other graphing utility to produce graphs of $f(x) = \log_2(x)$ and $h(x) = \sqrt[3]{x - 1}$ for $x \le 28$.

   b. Compare the graph of the function $f(x) = \log_2(x)$ with the graph of the function $h(x) = \sqrt[3]{x - 1}$. Describe the similarities and differences between the graphs.

   c. Is it always the case that $\log_2(x) > \sqrt[3]{x - 1}$ for $x > 2$?

This page intentionally left blank

Lesson 18: Graphs of Exponential Functions and Logarithmic Functions

Classwork

Opening Exercise

Complete the following table of values of the function $f(x) = 2^x$. We want to sketch the graph of $y = f(x)$ and then reflect that graph across the diagonal line with equation $y = x$.

x	$y = 2^x$	Point (x, y) on the graph of $y = 2^x$
-3		
-2		
-1		
0		
1		
2		
3		

On the set of axes below, plot the points from the table and sketch the graph of $y = 2^x$. Next, sketch the diagonal line with equation $y = x$, and then reflect the graph of $y = 2^x$ across the line.

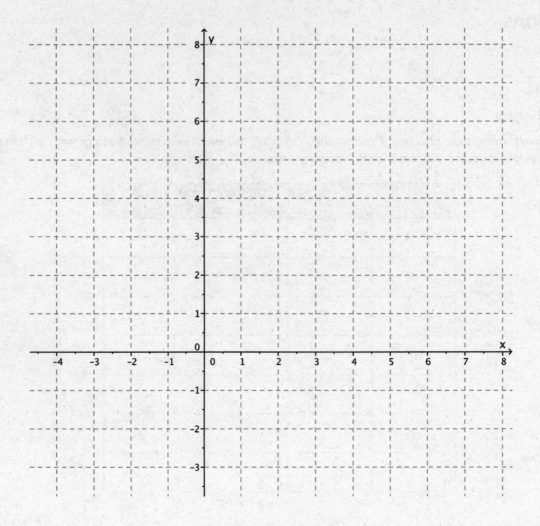

EUREKA
MATH™

©2015 Great Minds. eureka-math.org
ALG II-M3-SE-B2-1.3.0-08.2015

Exercises

1. Complete the following table of values of the function $g(x) = \log_2(x)$. We want to sketch the graph of $y = g(x)$ and then reflect that graph across the diagonal line with equation $y = x$.

x	$y = \log_2(x)$	Point (x, y) on the graph of $y = \log_2(x)$
$-\dfrac{1}{8}$		
$-\dfrac{1}{4}$		
$-\dfrac{1}{2}$		
1		
2		
4		
8		

On the set of axes below, plot the points from the table and sketch the graph of $y = \log_2(x)$. Next, sketch the diagonal line with equation $y = x$, and then reflect the graph of $y = \log_2(x)$ across the line.

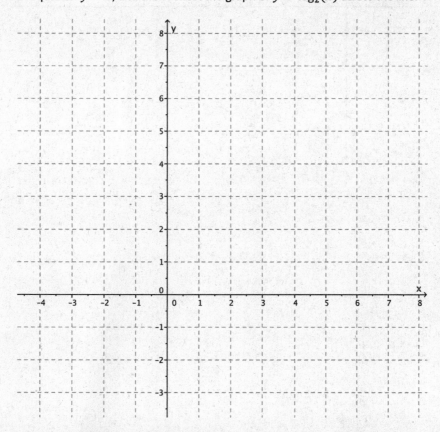

2. Working independently, predict the relation between the graphs of the functions $f(x) = 3^x$ and $g(x) = \log_3(x)$. Test your predictions by sketching the graphs of these two functions. Write your prediction in your notebook, provide justification for your prediction, and compare your prediction with that of your neighbor.

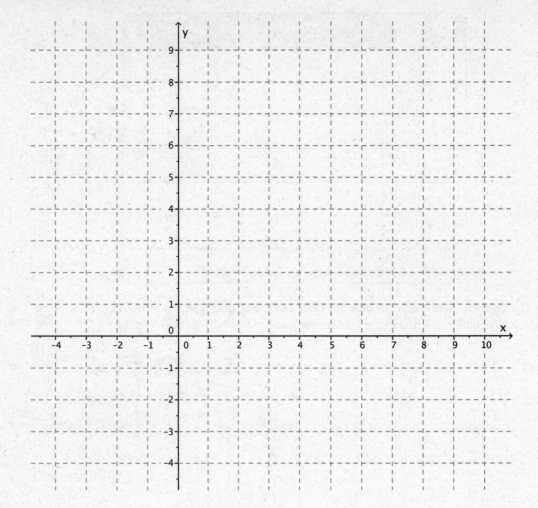

Lesson 18: Graphs of Exponential Functions and Logarithmic Functions

EUREKA
MATH™

3. Now let's compare the graphs of the functions $f_2(x) = 2^x$ and $f_3(x) = 3^x$. Sketch the graphs of the two exponential functions on the same set of axes; then, answer the questions below.

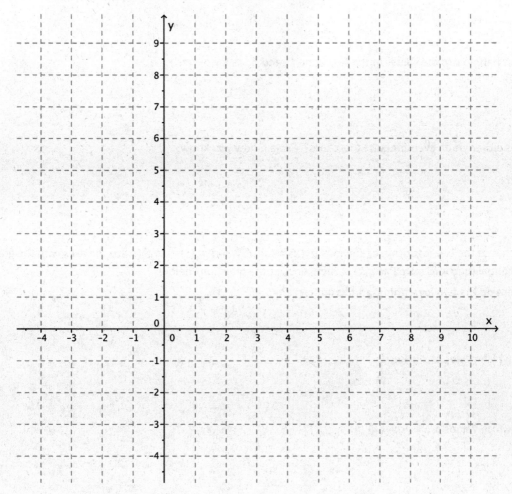

a. Where do the two graphs intersect?

b. For which values of x is $2^x < 3^x$?

c. For which values of x is $2^x > 3^x$?

d. What happens to the values of the functions f_2 and f_3 as $x \to \infty$?

e. What happens to the values of the functions f_2 and f_3 as $x \to -\infty$?

f. Does either graph ever intersect the x-axis? Explain how you know.

4. Add sketches of the two logarithmic functions $g_2(x) = \log_2(x)$ and $g_3(x) = \log_3(x)$ to the axes with the graphs of the exponential functions from Exercise 3; then, answer the questions below.

a. Where do the two logarithmic graphs intersect?

b. For which values of x is $\log_2(x) < \log_3(x)$?

c. For which values of x is $\log_2(x) > \log_3(x)$?

d. What happens to the values of the functions g_2 and g_3 as $x \to \infty$?

e. What happens to the values of the functions g_2 and g_3 as $x \to 0$?

f. Does either graph ever intersect the y-axis? Explain how you know.

g. Describe the similarities and differences in the behavior of $f_2(x)$ and $g_2(x)$ as $x \to \infty$.

©2015 Great Minds. eureka-math.org
ALG II-M3-SE-B2-1.3.0-08.2015

Problem Set

1. Sketch the graphs of the functions $f(x) = 5^x$ and $g(x) = \log_5(x)$.

2. Sketch the graphs of the functions $f(x) = \left(\frac{1}{2}\right)^x$ and $g(x) = \log_{\frac{1}{2}}(x)$.

3. Sketch the graphs of the functions $f_1(x) = \left(\frac{1}{2}\right)^x$ and $f_2(x) = \left(\frac{3}{4}\right)^x$ on the same sheet of graph paper and answer the following questions.

   a. Where do the two exponential graphs intersect?

   b. For which values of x is $\left(\frac{1}{2}\right)^x < \left(\frac{3}{4}\right)^x$?

   c. For which values of x is $\left(\frac{1}{2}\right)^x > \left(\frac{3}{4}\right)^x$?

   d. What happens to the values of the functions f_1 and f_2 as $x \to \infty$?

   e. What are the domains of the two functions f_1 and f_2?

4. Use the information from Problem 3 together with the relationship between graphs of exponential and logarithmic functions to sketch the graphs of the functions $g_1(x) = \log_{\frac{1}{2}}(x)$ and $g_2(x) = \log_{\frac{3}{4}}(x)$ on the same sheet of graph paper. Then, answer the following questions.

   a. Where do the two logarithmic graphs intersect?

   b. For which values of x is $\log_{\frac{1}{2}}(x) < \log_{\frac{3}{4}}(x)$?

   c. For which values of x is $\log_{\frac{1}{2}}(x) > \log_{\frac{3}{4}}(x)$?

   d. What happens to the values of the functions g_1 and g_2 as $x \to \infty$?

   e. What are the domains of the two functions g_1 and g_2?

5. For each function f, find a formula for the function h in terms of x.

   a. If $f(x) = x^3$, find $h(x) = 128f\left(\frac{1}{4}x\right) + f(2x)$.

   b. If $f(x) = x^2 + 1$, find $h(x) = f(x + 2) - f(2)$.

   c. If $f(x) = x^3 + 2x^2 + 5x + 1$, find $h(x) = \frac{f(x) + f(-x)}{2}$.

   d. If $f(x) = x^3 + 2x^2 + 5x + 1$, find $h(x) = \frac{f(x) - f(-x)}{2}$.

6. In Problem 5, parts (c) and (d), list at least two aspects about the formulas you found as they relate to the function $f(x) = x^3 + 2x^2 + 5x + 1$.

EUREKA
MATH™

Lesson 18:   Graphs of Exponential Functions and Logarithmic Functions

S.135

©2015 Great Minds. eureka-math.org
ALG II-M3-SE-B2-1.3.0-08.2015

7. For each of the functions f and g below, write an expression for (i) $f\big(g(x)\big)$, (ii) $g\big(f(x)\big)$, and (iii) $f\big(f(x)\big)$ in terms of x.

a. $f(x) = x^{\frac{2}{3}}$, $g(x) = x^{12}$

b. $f(x) = \frac{b}{x-a}$, $g(x) = \frac{b}{x} + a$ for two numbers a and b, when x is not equal to 0 or a

c. $f(x) = \frac{x+1}{x-1}$, $g(x) = \frac{x+1}{x-1}$, when x is not equal to 1 or -1

d. $f(x) = 2^x$, $g(x) = \log_2(x)$

e. $f(x) = \ln(x)$, $g(x) = e^x$

f. $f(x) = 2 \cdot 100^x$, $g(x) = \frac{1}{2}\log\left(\frac{1}{2}x\right)$

EUREKA
MATH™

Lesson 19: The Inverse Relationship Between Logarithmic and Exponential Functions

Classwork

Opening Exercise

a. Consider the mapping diagram of the function f below. Fill in the blanks of the mapping diagram of g to construct a function that undoes each output value of f by returning the original input value of f. (The first one is done for you.)

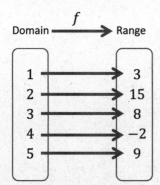 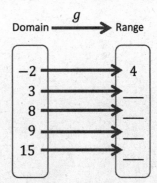

b. Write the set of input-output pairs for the functions f and g by filling in the blanks below. (The set F for the function f has been done for you.)

$F = \{(1,3), (2,15), (3,8), (4,-2), (5,9)\}$

$G = \{(-2,4), \underline{\hspace{1cm}}, \underline{\hspace{1cm}}, \underline{\hspace{1cm}}, \underline{\hspace{1cm}}\}$

c. How can the points in the set G be obtained from the points in F?

d. Peter studied the mapping diagrams of the functions f and g above and exclaimed, "I can get the mapping diagram for g by simply taking the mapping diagram for f and reversing all of the arrows!" Is he correct?

Exercises

For each function f in Exercises 1–5, find the formula for the corresponding inverse function g. Graph both functions on a calculator to check your work.

1. $f(x) = 1 - 4x$

2. $f(x) = x^3 - 3$

3. $f(x) = 3 \log(x^2)$ for $x > 0$

4. $f(x) = 2^{x-3}$

5. $f(x) = \dfrac{x+1}{x-1}$ for $x \neq 1$

6. Cindy thinks that the inverse of $f(x) = x - 2$ is $g(x) = 2 - x$. To justify her answer, she calculates $f(2) = 0$ and then substitutes the output 0 into g to get $g(0) = 2$, which gives back the original input. Show that Cindy is incorrect by using other examples from the domain and range of f.

7. After finding the inverse for several functions, Henry claims that every function must have an inverse. Rihanna says that his statement is not true and came up with the following example: If $f(x) = |x|$ has an inverse, then because $f(3)$ and $f(-3)$ both have the same output 3, the inverse function g would have to map 3 to both 3 and -3 simultaneously, which violates the definition of a function. What is another example of a function without an inverse?

Example

Consider the function $f(x) = 2^x + 1$, whose graph is shown to the right.

a. What are the domain and range of f?

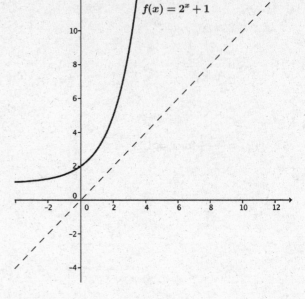

b. Sketch the graph of the inverse function g on the graph. What type of function do you expect g to be?

c. What are the domain and range of g? How does that relate to your answer in part (a)?

d. Find the formula for g.

EUREKA
MATH™

Lesson Summary

- **INVERTIBLE FUNCTION:** Let f be a function whose domain is the set X and whose image is the set Y. Then f is *invertible* if there exists a function g with domain Y and image X such that f and g satisfy the property:

  For all x in X and y in Y, $f(x) = y$ if and only if $g(y) = x$.

- The function g is called the *inverse* of f.

- If two functions whose domain and range are a subset of the real numbers are inverses, then their graphs are reflections of each other across the diagonal line given by $y = x$ in the Cartesian plane.

- If f and g are inverses of each other, then
  - The domain of f is the same set as the range of g.
  - The range of f is the same set as the domain of g.

- In general, to find the formula for an inverse function g of a given function f:
  - Write $y = f(x)$ using the formula for f.
  - Interchange the symbols x and y to get $x = f(y)$.
  - Solve the equation for y to write y as an expression in x.
  - Then, the formula for g is the expression in x found in step (iii).

- The functions $f(x) = \log_b(x)$ and $g(x) = b^x$ are inverses of each other.

Problem Set

1. For each function h below, find two functions f and g such that $h(x) = f\big(g(x)\big)$. (There are many correct answers.)

   a. $h(x) = (3x + 7)^2$

   b. $h(x) = \sqrt[3]{x^2 - 8}$

   c. $h(x) = \dfrac{1}{2x - 3}$

   d. $h(x) = \dfrac{4}{(2x - 3)^3}$

   e. $h(x) = (x + 1)^2 + 2(x + 1)$

   f. $h(x) = (x + 4)^{\frac{4}{5}}$

   g. $h(x) = \sqrt[3]{\log(x^2 + 1)}$

   h. $h(x) = \sin(x^2 + 2)$

   i. $h(x) = \ln(\sin(x))$

2. Let f be the function that assigns to each student in your class his or her biological mother.

   a. Use the definition of function to explain why f is a function.

   b. In order for f to have an inverse, what condition must be true about the students in your class?

   c. If we enlarged the domain to include all students in your school, would this larger domain function have an inverse?

3. The table below shows a partially filled-out set of input-output pairs for two functions f and h that have the same finite domain of $\{0, 5, 10, 15, 20, 25, 30, 35, 40\}$.

x	0	5	10	15	20	25	30	35	40
$f(x)$	0	0.3	1.4		2.1		2.7	6	
$h(x)$	0	0.3	1.4		2.1		2.7	6	

   a. Complete the table so that f is invertible but h is definitely not invertible.

   b. Graph both functions and use their graphs to explain why f is invertible and h is not.

4. Find the inverse of each of the following functions. In each case, indicate the domain and range of both the original function and its inverse.

   a. $f(x) = \dfrac{3x - 7}{5}$

   b. $f(x) = \dfrac{5 + x}{6 - 2x}$

   c. $f(x) = e^{x-5}$

   d. $f(x) = 2^{5-8x}$

   e. $f(x) = 7\log(1 + 9x)$

   f. $f(x) = 8 + \ln(5 + \sqrt[3]{x})$

   g. $f(x) = \log\left(\dfrac{100}{3x+2}\right)$

   h. $f(x) = \ln(x) - \ln(x + 1)$

   i. $f(x) = \dfrac{2^x}{2^x + 1}$

5. Even though there are no real principal square roots for negative numbers, principal cube roots do exist for negative numbers: $\sqrt[3]{-8}$ is the real number -2 since $-2 \cdot -2 \cdot -2 = -8$. Use the identities $\sqrt[3]{x^3} = x$ and $\left(\sqrt[3]{x}\right)^3 = x$ for any real number x to find the inverse of each of the functions below. In each case, indicate the domain and range of both the original function and its inverse.

   a. $f(x) = \sqrt[3]{2x}$ for any real number x.

   b. $f(x) = \sqrt[3]{2x - 3}$ for any real number x.

   c. $f(x) = (x - 1)^3 + 3$ for any real number x.

6. Suppose that the inverse of a function is the function itself. For example, the inverse of the function $f(x) = \frac{1}{x}$ (for $x \neq 0$) is just itself again, $g(x) = \frac{1}{x}$ (for $x \neq 0$). What symmetry must the graphs of all such functions have? (Hint: Study the graph of Exercise 5 in the lesson.)

7. There are two primary scales for measuring daily temperature: degrees Celsius and degrees Fahrenheit. The United States uses the Fahrenheit scale, and many other countries use the Celsius scale. When traveling abroad you will often need to convert between these two temperature scales.

Let f be the function that inputs a temperature measure in degrees Celsius, denoted by °C, and outputs the corresponding temperature measure in degrees Fahrenheit, denoted by °F.

   a. Assuming that f is linear, we can use two points on the graph of f to determine a formula for f. In degrees Celsius, the freezing point of water is 0, and its boiling point is 100. In degrees Fahrenheit, the freezing point of water is 32, and its boiling point is 212. Use this information to find a formula for the function f. (Hint: Plot the points and draw the graph of f first, keeping careful track of the meaning of values on the x-axis and y-axis.)

   b. If the temperature in Paris is 25°C, what is the temperature in degrees Fahrenheit?

   c. Find the inverse of the function f and explain its meaning in terms of degrees Fahrenheit and degrees Celsius.

   d. The graphs of f and its inverse g are two lines that intersect in one point. What is that point? What is its significance in terms of degrees Celsius and degrees Fahrenheit?

Extension: Use the fact that, for $b > 1$, the functions $f(x) = b^x$ and $g(x) = \log_b(x)$ are increasing to solve the following problems. Recall that an increasing function f has the property that if both a and b are in the domain of f and $a < b$, then $f(a) < f(b)$.

8. For which values of x is $2^x < \frac{1}{1,000,000}$?

9. For which values of x is $\log_2(x) < -1,000,000$?

This page intentionally left blank

Lesson 20: Transformations of the Graphs of Logarithmic and Exponential Functions

Classwork

Opening Exercise

a. Sketch the graphs of the three functions $f(x) = x^2$, $g(x) = (2x)^2 + 1$, and $h(x) = 4x^2 + 1$.

   i. Describe the sequence of transformations that takes the graph of $f(x) = x^2$ to the graph of $g(x) = (2x)^2 + 1$.

   ii. Describe the sequence of transformations that takes the graph of $f(x) = x^2$ to the graph of $h(x) = 4x^2 + 1$.

   iii. Explain why g and h from parts (i) and (ii) are equivalent functions.

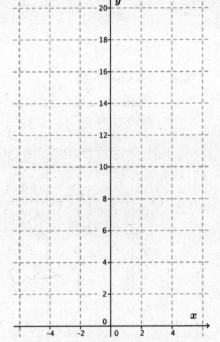

b. Describe the sequence of transformations that takes the graph of $f(x) = \sin(x)$ to the graph of $g(x) = \sin(2x) - 3$.

EUREKA
MATH™

©2015 Great Minds. eureka-math.org
ALG II-M3-SE-B2-1.3.0-08.2015

c. Describe the sequence of transformations that takes the graph of $f(x) = \sin(x)$ to the graph of $h(x) = 4\sin(x) - 3$.

d. Explain why g and h from parts (b)–(c) are *not* equivalent functions.

Exploratory Challenge

a. Sketch the graph of $f(x) = \log_2(x)$ by identifying and plotting at least five key points. Use the table below to get started.

x	$\log_2(x)$
$\dfrac{1}{4}$	-2
$\dfrac{1}{2}$	-1
1	
2	
4	
8	

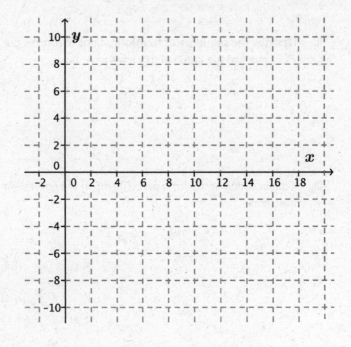

Lesson 20:   Transformations of the Graphs of Logarithmic and Exponential Functions

EUREKA
MATH™

©2015 Great Minds. eureka-math.org
ALG II-M3-SE-B2-1.3.0-08.2015

b. Describe a sequence of transformations that takes the graph of f to the graph of $g(x) = \log_2(8x)$.

c. Describe a sequence of transformations that takes the graph of f to the graph of $h(x) = 3 + \log_2(x)$.

d. Complete the table below for f, g, and h and describe any noticeable patterns.

x	$f(x)$	$g(x)$	$h(x)$
$\dfrac{1}{4}$			
$\dfrac{1}{2}$			
1			
2			
4			
8			

e. Graph the three functions on the same coordinate axes and describe any noticeable patterns.

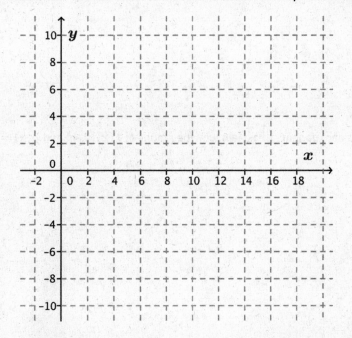

EUREKA
MATH™

Lesson 20: Transformations of the Graphs of Logarithmic and Exponential Functions

S.147

©2015 Great Minds. eureka-math.org
ALG II-M3-SE-B2-1.3.0-08.2015

f. Use a property of logarithms to show that g and h are equivalent.

g. Describe the graph of $p(x) = \log_2\left(\frac{x}{4}\right)$ as a vertical translation of the graph of $f(x) = \log_2(x)$. Justify your response.

h. Describe the graph of $h(x) = 3 + \log_2(x)$ as a horizontal scaling of the graph of $f(x) = \log_2(x)$. Justify your response.

i. Do the functions $h(x) = \log_2(8) + \log_2(x)$ and $k(x) = \log_2(x + 8)$ have the same graphs? Justify your reasoning.

j. Use the properties of exponents to explain why the graphs of $f(x) = 4^x$ and $g(x) = 2^{2x}$ are identical.

EUREKA
MATH™

©2015 Great Minds. eureka-math.org
ALG II-M3-SE-B2-1.3.0-08.2015

k. Use the properties of exponents to predict what the graphs of $g(x) = 4 \cdot 2^x$ and $h(x) = 2^{x+2}$ look like compared to one another. Describe the graphs of g and h as transformations of the graph of $f(x) = 2^x$. Confirm your prediction by graphing f, g, and h on the same coordinate axes.

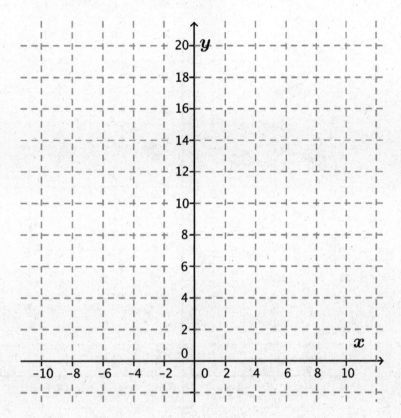

I. Graph $f(x) = 2^x$, $g(x) = 2^{-x}$, and $h(x) = \left(\frac{1}{2}\right)^x$ on the same coordinate axes. Describe the graphs of g and h as transformations of the graph of f. Use the properties of exponents to explain why g and h are equivalent.

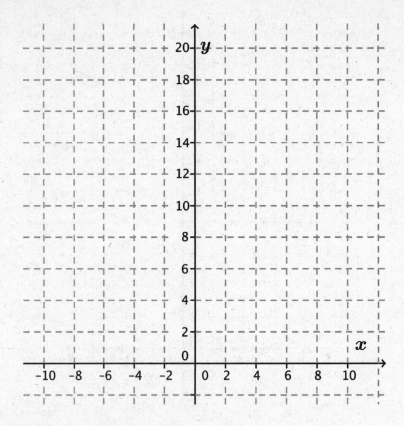

Example 1: Graphing Transformations of the Logarithm Functions

The general form of a logarithm function is given by $f(x) = k + a \log_b(x - h)$, where a, b, k, and h are real numbers such that b is a positive number not equal to 1 and $x - h > 0$.

a. Given $g(x) = 3 + 2 \log(x - 2)$, describe the graph of g as a transformation of the common logarithm function.

EUREKA
MATH

b. Graph the common logarithm function and g on the same coordinate axes.

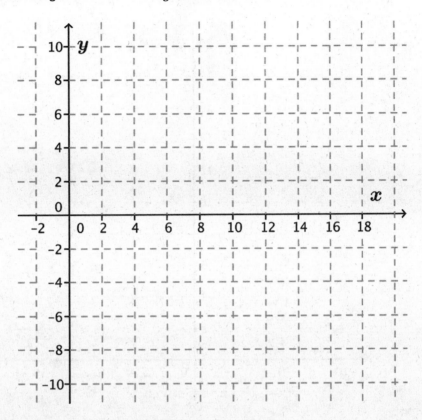

Example 2: Graphing Transformations of Exponential Functions

The general form of the exponential function is given by $f(x) = a \cdot b^x + k$, where a, b, and k are real numbers such that b is a positive number not equal to 1.

a. Use the properties of exponents to transform the function $g(x) = 3^{2x+1} - 2$ to the general form, and then graph it. What are the values of a, b, and k?

b. Describe the graph of g as a transformation of the graph of $f(x) = 9^x$.

c. Describe the graph of g as a transformation of the graph of $f(x) = 3^x$.

d.   Sketch the graph of g using transformations.

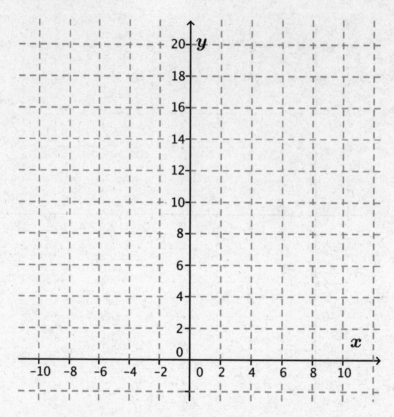

Exercises

Graph each pair of functions by first graphing f and then graphing g by applying transformations of the graph of f. Describe the graph of g as a transformation of the graph of f.

1. $f(x) = \log_3(x)$ and $g(x) = 2\log_3(x - 1)$

2. $f(x) = \log(x)$ and $g(x) = \log(100x)$

3. $f(x) = \log_5 x$ and $g(x) = -\log_5\big(5(x + 2)\big)$

4. $f(x) = 3^x$ and $g(x) = -2 \cdot 3^{x-1}$

Lesson Summary

GENERAL FORM OF A LOGARITHMIC FUNCTION: $f(x) = k + a \log_b(x - h)$ such that a, h, and k are real numbers, b is any positive number not equal to 1, and $x - h > 0$.

GENERAL FORM OF AN EXPONENTIAL FUNCTION: $f(x) = a \cdot b^x + k$ such that a and k are real numbers, and b is any positive number not equal to 1.

The properties of logarithms and exponents can be used to rewrite expressions for functions in equivalent forms that can then be graphed by applying transformations.

Problem Set

1. Describe each function as a transformation of the graph of a function in the form $f(x) = \log_b(x)$. Sketch the graph of f and the graph of g by hand. Label key features such as intercepts, intervals where g is increasing or decreasing, and the equation of the vertical asymptote.

   a. $g(x) = \log_2(x - 3)$

   b. $g(x) = \log_2(16x)$

   c. $g(x) = \log_2\left(\dfrac{8}{x}\right)$

   d. $g(x) = \log_2((x - 3)^2)$ for $x > 3$

2. Each function graphed below can be expressed as a transformation of the graph of $f(x) = \log(x)$. Write an algebraic function for g and h and state the domain and range.

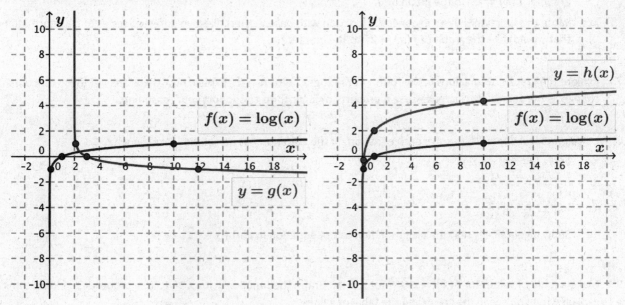

Figure 1: Graphs of $f(x) = \log(x)$ and the function g **Figure 2:** Graphs of $f(x) = \log(x)$ and the function h

3. Describe each function as a transformation of the graph of a function in the form $f(x) = b^x$. Sketch the graph of f and the graph of g by hand. Label key features such as intercepts, intervals where g is increasing or decreasing, and the horizontal asymptote.

   a. $g(x) = 2 \cdot 3^x - 1$

   b. $g(x) = 2^{2x} + 3$

   c. $g(x) = 3^{x-2}$

   d. $g(x) = -9^{\frac{x}{2}} + 1$

4. Using the function $f(x) = 2^x$, create a new function g whose graph is a series of transformations of the graph of f with the following characteristics:

   ▪ The function g is decreasing for all real numbers.

   ▪ The equation for the horizontal asymptote is $y = 5$.

   ▪ The y-intercept is 7.

5. Using the function $f(x) = 2^x$, create a new function g whose graph is a series of transformations of the graph of f with the following characteristics:

   ▪ The function g is increasing for all real numbers.

   ▪ The equation for the horizontal asymptote is $y = 5$.

   ▪ The y-intercept is 4.

6. Consider the function $g(x) = \left(\frac{1}{4}\right)^{x-3}$:

   a. Write the function g as an exponential function with base 4. Describe the transformations that would take the graph of $f(x) = 4^x$ to the graph of g.

   b. Write the function g as an exponential function with base 2. Describe two different series of transformations that would take the graph of $f(x) = 2^x$ to the graph of g.

7. Explore the graphs of functions in the form $f(x) = \log(x^n)$ for $n > 1$. Explain how the graphs of these functions change as the values of n increase. Use a property of logarithms to support your reasoning.

8. Use a graphical approach to solve each equation. If the equation has no solution, explain why.

   a. $\log(x) = \log(x - 2)$

   b. $\log(x) = \log(2x)$

   c. $\log(x) = \log\left(\frac{2}{x}\right)$

   d. Show algebraically that the exact solution to the equation in part (c) is $\sqrt{2}$.

9. Make a table of values for $f(x) = x^{\frac{1}{\log(x)}}$ for $x > 1$. Graph the function f for $x > 1$. Use properties of logarithms to explain what you see in the graph and the table of values.

EUREKA
MATH™

Lesson 21: The Graph of the Natural Logarithm Function

Classwork

Exploratory Challenge

Your task is to compare graphs of base b logarithm functions to the graph of the common logarithm function $f(x) = \log(x)$ and summarize your results with your group. Recall that the base of the common logarithm function is 10. A graph of f is provided below.

a.   Select at least one base value from this list: $\frac{1}{10}, \frac{1}{2}, 2, 5, 20, 100$. Write a function in the form $g(x) = \log_b(x)$ for your selected base value, b.

b.   Graph the functions f and g in the same viewing window using a graphing calculator or other graphing application, and then add a sketch of the graph of g to the graph of f shown below.

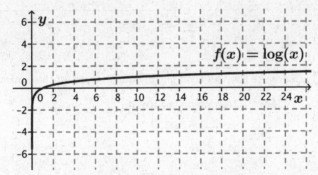

c.   Describe how the graph of g for the base you selected compares to the graph of $f(x) = \log(x)$.

©2015 Great Minds. eureka-math.org
ALG II-M3-SE-B2-1.3.0-08.2015

d.   Share your results with your group and record observations on the graphic organizer below. Prepare a group presentation that summarizes the group's findings.

How does the graph of $g(x) = \log_b(x)$ compare to the graph of $f(x) = \log(x)$ for various values of b?	
$0 < b < 1$	
$1 < b < 10$	
$b > 10$	

Exercise 1

Use the change of base property to rewrite each logarithmic function in terms of the common logarithm function.

Base b Base 10 (Common Logarithm)

$g_1(x) = \log_{\frac{1}{4}}(x)$

$g_2(x) = \log_{\frac{1}{2}}(x)$

$g_3(x) = \log_2(x)$

$g_4(x) = \log_5(x)$

$g_5(x) = \log_{20}(x)$

$g_6(x) = \log_{100}(x)$

Example 1: The Graph of the Natural Logarithm Function $f(x) = \ln(x)$

Graph the natural logarithm function below to demonstrate where it sits in relation to the graphs of the base-2 and base-10 logarithm functions.

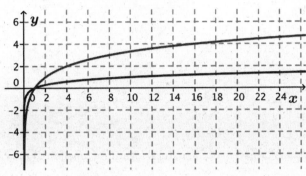

Example 2

Graph each function by applying transformations of the graphs of the natural logarithm function.

a. $f(x) = 3\ln(x - 1)$

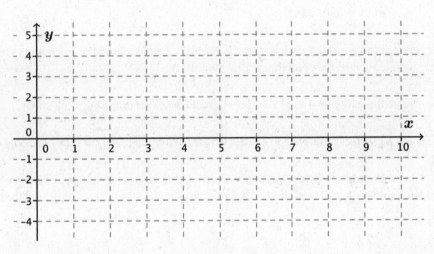

b. $g(x) = \log_6(x) - 2$

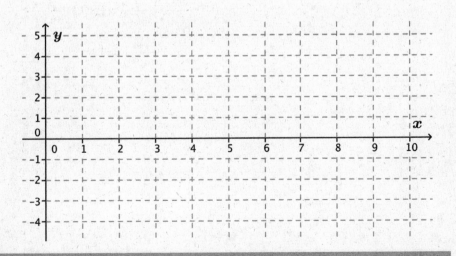

©2015 Great Minds. eureka-math.org
ALG II-M3-SE-B2-1.3.0-08.2015

Problem Set

1. Rewrite each logarithmic function as a natural logarithm function.

   a. $f(x) = \log_5(x)$

   b. $f(x) = \log_2(x - 3)$

   c. $f(x) = \log_2\left(\frac{x}{3}\right)$

   d. $f(x) = 3 - \log(x)$

   e. $f(x) = 2\log(x + 3)$

   f. $f(x) = \log_5(25x)$

2. Describe each function as a transformation of the natural logarithm function $f(x) = \ln(x)$.

   a. $g(x) = 3\ln(x + 2)$

   b. $g(x) = -\ln(1 - x)$

   c. $g(x) = 2 + \ln(e^2 x)$

   d. $g(x) = \log_5(25x)$

3. Sketch the graphs of each function in Problem 2 and identify the key features including intercepts, decreasing or increasing intervals, and the vertical asymptote.

4. Solve the equation $1 - e^{x-1} = \ln(x)$ graphically, without using a calculator.

5. Use a graphical approach to explain why the equation $\log(x) = \ln(x)$ has only one solution.

6. Juliet tried to solve this equation as shown below using the change of base property and concluded there is no solution because $\ln(10) \neq 1$. Construct an argument to support or refute her reasoning.

$$\log(x) = \ln(x)$$

$$\frac{\ln(x)}{\ln(10)} = \ln(x)$$

$$\left(\frac{\ln(x)}{\ln(10)}\right)\frac{1}{\ln(x)} = (\ln(x))\frac{1}{\ln(x)}$$

$$\frac{1}{\ln(10)} = 1$$

©2015 Great Minds. eureka-math.org
ALG II-M3-SE-B2-1.3.0-08.2015

7. Consider the function f given by $f(x) = \log_x(100)$ for $x > 0$ and $x \neq 1$.

    a. What are the values of $f(100)$, $f(10)$, and $f(\sqrt{10})$?

    b. Why is the value 1 excluded from the domain of this function?

    c. Find a value x so that $f(x) = 0.5$.

    d. Find a value w so that $f(w) = -1$.

    e. Sketch a graph of $y = \log_x(100)$ for $x > 0$ and $x \neq 1$.

©2015 Great Minds. eureka-math.org
ALG II-M3-SE-B2-1.3.0-08.2015

This page intentionally left blank

Lesson 22: Choosing a Model

Classwork

Opening Exercise

a. You are working on a team analyzing the following data gathered by your colleagues:

$$(-1.1, 5), (0, 105), (1.5, 178), (4.3, 120)$$

Your coworker Alexandra says that the model you should use to fit the data is

$$k(t) = 100 \cdot \sin(1.5t) + 105.$$

Sketch Alexandra's model on the axes at left on the next page.

b. How does the graph of Alexandra's model $k(t) = 100 \cdot \sin(1.5t) + 105$ relate to the four points? Is her model a good fit to this data?

c. Another teammate Randall says that the model you should use to fit the data is

$$g(t) = -16t^2 + 72t + 105.$$

Sketch Randall's model on the axes at right on the next page.

d. How does the graph of Randall's model $g(t) = -16t^2 + 72t + 105$ relate to the four points? Is his model a good fit to the data?

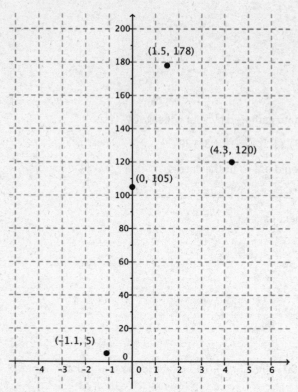

Alexandra's Model

Randall's Model

e. Suppose the four points represent positions of a projectile fired into the air. Which of the two models is more appropriate in that situation, and why?

f. In general, how do we know which model to choose?

Exercises

1. The table below contains the number of daylight hours in Oslo, Norway, on the specified dates.

Date	Hours and Minutes	Hours
August 1	16:56	16.93
September 1	14:15	14.25
October 1	11:33	11.55
November 1	8:50	8.83

a. Plot the data on the grid provided and decide how to best represent it.

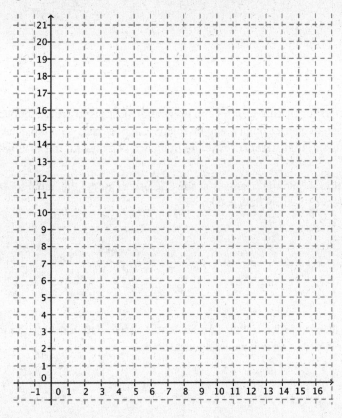

b. Looking at the data, what type of function appears to be the best fit?

c. Looking at the context in which the data was gathered, what type of function should be used to model the data?

d. Do you have enough information to find a model that is appropriate for this data? Either find a model or explain what other information you would need to do so.

2. The goal of the U.S. Centers for Disease Control and Prevention (CDC) is to protect public health and safety through the control and prevention of disease, injury, and disability. Suppose that 45 people have been diagnosed with a new strain of the flu virus and that scientists estimate that each person with the virus will infect 5 people every day with the flu.

a. What type of function should the scientists at the CDC use to model the initial spread of this strain of flu to try to prevent an epidemic? Explain how you know.

b. Do you have enough information to find a model that is appropriate for this situation? Either find a model or explain what other information you would need to do so.

EUREKA
MATH™

3. An artist is designing posters for a new advertising campaign. The first poster takes 10 hours to design, but each subsequent poster takes roughly 15 minutes less time than the previous one as he gets more practice.

    a. What type of function models the amount of time needed to create n posters, for $n \leq 20$? Explain how you know.

    b. Do you have enough information to find a model that is appropriate for this situation? Either find a model or explain what other information you would need to do so.

4. A homeowner notices that her heating bill is the lowest in the month of August and increases until it reaches its highest amount in the month of February. After February, the amount of the heating bill slowly drops back to the level it was in August, when it begins to increase again. The amount of the bill in February is roughly four times the amount of the bill in August.

    a. What type of function models the amount of the heating bill in a particular month? Explain how you know.

    b. Do you have enough information to find a model that is appropriate for this situation? Either find a model or explain what other information you would need to do so.

EUREKA
MATH™

Lesson 22:   Choosing a Model

S.165

©2015 Great Minds. eureka-math.org
ALG II-M3-SE-B2-1.3.0-08.2015

5. An online merchant sells used books for $5.00 each, and the sales tax rate is 6% of the cost of the books. Shipping charges are a flat rate of $4.00 plus an additional $1.00 per book.

   a. What type of function models the total cost, including the shipping costs, of a purchase of x books? Explain how you know.

   b. Do you have enough information to find a model that is appropriate for this situation? Either find a model or explain what other information you would need to do so.

6. A stunt woman falls from a tall building in an action-packed movie scene. Her speed increases by 32 ft/s for every second that she is falling.

   a. What type of function models her distance from the ground at time t seconds? Explain how you know.

   b. Do you have enough information to find a model that is appropriate for this situation? Either find a model or explain what other information you would need to do so.

©2015 Great Minds. eureka-math.org
ALG II-M3-SE-B2-1.3.0-08.2015

Lesson Summary

- If we expect from the context that each new term in the sequence of data is a constant added to the previous term, then we try a linear model.

- If we expect from the context that the second differences of the sequence are constant (meaning that the rate of change between terms either grows or shrinks linearly), then we try a quadratic model.

- If we expect from the context that each new term in the sequence of data is a constant multiple of the previous term, then we try an exponential model.

- If we expect from the context that the sequence of terms is periodic, then we try a sinusoidal model.

Model	Equation of Function	Rate of Change
Linear	$f(t) = at + b$ for $a \neq 0$	Constant
Quadratic	$g(t) = at^2 + bt + c$ for $a \neq 0$	Changing linearly
Exponential	$h(t) = ab^{ct}$ for $0 < b < 1$ or $b > 1$	A multiple of the current value
Sinusoidal	$k(t) = A\sin\big(w(t - h)\big) + k$ for $A, w \neq 0$	Periodic

Problem Set

1. A new car depreciates at a rate of about 20% per year, meaning that its resale value decreases by roughly 20% each year. After hearing this, Brett said that if you buy a new car this year, then after 5 years the car has a resale value of $0.00. Is his reasoning correct? Explain how you know.

2. Alexei just moved to Seattle, and he keeps track of the average rainfall for a few months to see if the city deserves its reputation as the rainiest city in the United States.

Month	Average rainfall
July	0.93 in.
September	1.61 in.
October	3.24 in.
December	6.06 in.

What type of function should Alexei use to model the average rainfall in month t?

3. Sunny, who wears her hair long and straight, cuts her hair once per year on January 1, always to the same length. Her hair grows at a constant rate of 2 cm per month. Is it appropriate to model the length of her hair with a sinusoidal function? Explain how you know.

4. On average, it takes 2 minutes for a customer to order and pay for a cup of coffee.

   a. What type of function models the amount of time you wait in line as a function of how many people are in front of you? Explain how you know.

   b. Find a model that is appropriate for this situation.

5. An online ticket-selling service charges $50.00 for each ticket to an upcoming concert. In addition, the buyer must pay 8% sales tax and a convenience fee of $6.00 for the purchase.

   a. What type of function models the total cost of the purchase of n tickets in a single transaction?

   b. Find a model that is appropriate for this situation.

6. In a video game, the player must earn enough points to pass one level and progress to the next as shown in the table below.

To pass this level...	You need this many total points...
1	5,000
2	15,000
3	35,000
4	65,000

   That is, the increase in the required number of points increases by 10,000 points at each level.

   a. What type of function models the total number of points you need to pass to level n? Explain how you know.

   b. Find a model that is appropriate for this situation.

7. The southern white rhinoceros reproduces roughly once every three years, giving birth to one calf each time. Suppose that a nature preserve houses 100 white rhinoceroses, 50 of which are female. Assume that half of the calves born are female and that females can reproduce as soon as they are 1 year old.

   a. What type of function should be used to model the population of female white rhinoceroses in the preserve?

   b. Assuming that there is no death in the rhinoceros population, find a function to model the population of female white rhinoceroses in the preserve.

   c. Realistically, not all of the rhinoceroses survive each year, so we assume a 5% death rate of all rhinoceroses. Now what type of function should be used to model the population of female white rhinoceroses in the preserve?

   d. Find a function to model the population of female white rhinoceroses in the preserve, taking into account the births of new calves and the 5% death rate.

Lesson 23: Bean Counting

Classwork

Mathematical Modeling Exercises

1. Working with a partner, you are going to gather some data, analyze it, and find a function to use to model it. Be prepared to justify your choice of function to the class.

   a. Gather your data: For each trial, roll the beans from the cup to the paper plate. Count the number of beans that land marked side up, and add that many beans to the cup. Record the data in the table below. Continue until you have either completed 10 trials or the number of beans at the start of the trial exceeds the number that you have.

Trial Number, t	Number of Beans at Start of Trial	Number of Beans That Landed Marked-Side Up
1	1	
2		
3		
4		
5		
6		
7		
8		
9		
10		

   b. Based on the context in which you gathered this data, which type of function would best model your data points?

c. Plot the data: Plot the trial number on the horizontal axis and the number of beans in the cup at the start of the trial on the vertical axis. Be sure to label the axes appropriately and to choose a reasonable scale for the axes.

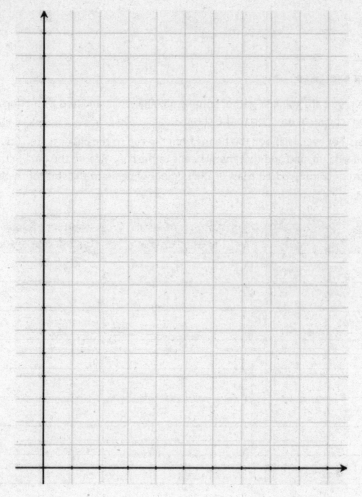

d. Analyze the data: Which type of function best fits your data? Explain your reasoning.

e. Model the data: Enter the data into the calculator and use the appropriate type of regression to find an equation that fits this data. Round the constants to two decimal places.

©2015 Great Minds. eureka-math.org
ALG II-M3-SE-B2-1.3.0-08.2015

2. This time, we are going to start with 50 beans in your cup. Roll the beans onto the plate and remove any beans that land marked-side up. Repeat until you have no beans remaining.

   a. Gather your data: For each trial, roll the beans from the cup to the paper plate. Count the number of beans that land marked-side up, and remove that many beans from the plate. Record the data in the table below. Repeat until you have no beans remaining.

Trial Number, t	Number of Beans at Start of Trial	Number of Beans That Landed Marked-Side Up
1	50	
2		
3		
4		
5		
6		
7		
8		
9		
10		

b.   Plot the data: Plot the trial number on the horizontal axis and the number of beans in the cup at the start of the trial on the vertical axis. Be sure to label the axes appropriately and choose a reasonable scale for the axes.

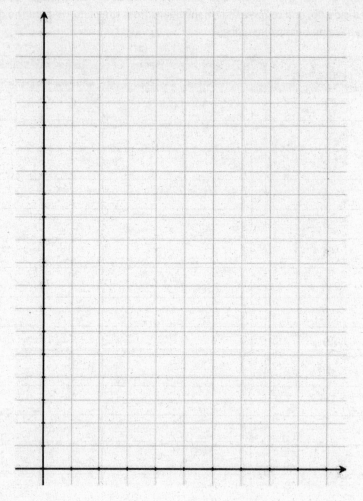

c.   Analyze the data: Which type of function best fits your data? Explain your reasoning.

d.   Make a prediction: What do you expect the values of a and b to be for your function? Explain your reasoning.

EUREKA
MATH™

e. Model the data: Enter the data into the calculator. Do not enter your final data point of 0 beans. Use the appropriate type of regression to find an equation that fits this data. Round the constants to two decimal places.

Problem Set

1. For this exercise, we consider three scenarios for which data have been collected and functions have been found to model the data, where $a, b, c, d, p, q, r, s, t,$ and u are positive real number constants.

   (i) The function $f(t) = a \cdot b^t$ models the original bean activity (Mathematical Modeling Exercise 1). Each bean is painted or marked on one side, and we start with one bean in the cup. A trial consists of throwing the beans in the cup and adding one more bean for each bean that lands marked side up.

   (ii) The function $g(t) = c \cdot d^t$ models a modified bean activity. Each bean is painted or marked on one side, and we start with one bean in the cup. A trial consists of throwing the beans in the cup and adding two more beans for each bean that lands marked side up.

   (iii) The function $h(t) = p \cdot q^t$ models the dice activity from the Exit Ticket. Start with one six-sided die in the cup. A trial consists of rolling the dice in the cup and adding one more die to the cup for each die that lands with a 6 showing.

   (iv) The function $j(t) = r \cdot s^t$ models a modified dice activity. Start with one six-sided die in the cup. A trial consists of rolling the dice in the cup and adding one more die to the cup for each die that lands with a 5 or a 6 showing.

   (v) The function $k(t) = u \cdot v^t$ models a modified dice activity. Start with one six-sided die in the cup. A trial consists of rolling the dice in the cup and adding one more die to the cup for each die that lands with an even number showing.

   a. What values do you expect for $a, c, p, r,$ and u?

   b. What value do you expect for the base b in the function $f(t) = a \cdot b^t$ in scenario (i)?

   c. What value do you expect for the base d in the function $g(t) = c \cdot d^t$ in scenario (ii)?

   d. What value do you expect for the base q in the function $h(t) = p \cdot q^t$ in scenario (iii)?

   e. What value do you expect for the base s in the function $j(t) = r \cdot s^t$ in scenario (iv)?

   f. What value do you expect for the base v in the function $k(t) = u \cdot v^t$ in scenario (v)?

   g. The following graphs represent the four functions $f, g, h,$ and j. Identify which graph represents which function.

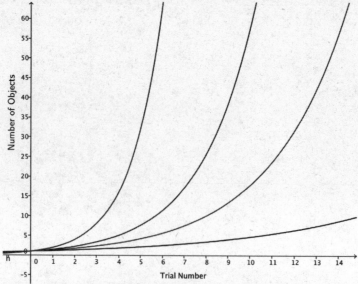

©2015 Great Minds. eureka-math.org
ALG II-M3-SE-B2-1.3.0-08.2015

2. Teams 1, 2, and 3 gathered data as shown in the tables below, and each team modeled their data using an exponential function of the form $f(t) = a \cdot b^t$.

   a. Which team should have the highest value of b? Which team should have the lowest value of b? Explain how you know.

Team 1		Team 2		Team 3	
Trial Number, t	Number of Beans	Trial Number, t	Number of Beans	Trial Number, t	Number of Beans
0	1	0	1	0	1
1	1	1	1	1	2
2	2	2	1	2	3
3	2	3	2	3	5
4	4	4	2	4	8
5	6	5	3	5	14
6	8	6	5	6	26
7	14	7	7	7	46
8	22	8	12	8	76
9	41	9	18	9	
10	59	10	27	10	

   b. Use a graphing calculator to find the equation that best fits each set of data. Do the equations of the functions provide evidence that your answer in part (a) is correct?

3. Omar has devised an activity in which he starts with 15 dice in his cup. A trial consists of rolling the dice in the cup and adding one more die to the cup for each die that lands with a 1, 2, or 3 showing.

   a. Find a function $f(t) = a(b^t)$ that Omar would expect to model his data.

   b. Solve the equation $f(t) = 30$. What does the solution mean?

   c. Omar wants to know in advance how many trials it should take for his initial quantity of 15 dice to double. He uses properties of exponents and logarithms to rewrite the function from part (a) as the exponential function $f(t) = 15\left(2^{t \log_2\left(\frac{3}{2}\right)}\right)$. Has Omar correctly applied the properties of exponents and logarithms to obtain an equivalent expression for his original equation in part (a)? Explain how you know.

   d. Explain how the modified formula from part (c) allows Omar to easily find the expected amount of time, t, for the initial quantity of dice to double.

4. Brenna has devised an activity in which she starts with 10 dice in her cup. A trial consists of rolling the dice in the cup and adding one more die to the cup for each die that lands with a 6 showing.

    a. Find a function $f(t) = a(b^t)$ that you would expect to model her data.

    b. Solve the equation $f(t) = 30$. What does your solution mean?

    c. Brenna wants to know in advance how many trials it should take for her initial quantity of 10 dice to triple. Use properties of exponents and logarithms to rewrite your function from part (a) as an exponential function of the form $f(t) = a(3^{ct})$.

    d. Explain how your formula from part (c) allows you to easily find the expected amount of time, t, for the initial quantity of dice to triple.

    e. Rewrite the formula for the function f using a base-10 exponential function.

    f. Use your formula from part (e) to find out how many trials it should take for the quantity of dice to grow to 100 dice.

5. Suppose that one bacteria population can be modeled by the function $P_1(t) = 500(2^t)$ and a second bacteria population can be modeled by the function $P_2(t) = 500(2.83^t)$, where t measures time in hours. Keep four digits of accuracy for decimal approximations of logarithmic values.

    a. What does the 500 mean in each function?

    b. Which population should double first? Explain how you know.

    c. How many hours and minutes should it take for the first population to double?

    d. Rewrite the formula for $P_2(t)$ in the form $P_2(t) = a(2^{ct})$, for some real numbers a and c.

    e. Use your formula in part (d) to find the time, t, in hours and minutes until the second population doubles.

6. Copper has antibacterial properties, and it has been shown that direct contact with copper alloy C11000 at 20°C kills 99.9% of all methicillin-resistant *Staphylococcus aureus* (MRSA) bacteria in about 75 minutes. Keep four digits of accuracy for decimal approximations of logarithmic values.

    a. A function that models a population of 1,000 MRSA bacteria t minutes after coming in contact with copper alloy C11000 is $P(t) = 1000(0.912)^t$. What does the base 0.912 mean in this scenario?

    b. Rewrite the formula for P as an exponential function with base $\frac{1}{2}$.

    c. Explain how your formula from part (b) allows you to easily find the time it takes for the population of MRSA to be reduced by half.

Lesson 24: Solving Exponential Equations

Classwork

Opening Exercise

In Lesson 7, we modeled a population of bacteria that doubled every day by the function $P(t) = 2^t$, where t was the time in days. We wanted to know the value of t when there were 10 bacteria. Since we didn't yet know about logarithms, we approximated the value of t numerically and we found that $P(t) = 10$ when $t \approx 3.32$.

Use your knowledge of logarithms to find an exact value for t when $P(t) = 10$, and then use your calculator to approximate that value to four decimal places.

Exercises

1. Fiona modeled her data from the bean-flipping experiment in Lesson 23 by the function $f(t) = 1.263(1.357)^t$, and Gregor modeled his data with the function $g(t) = 0.972(1.629)^t$.

    a. Without doing any calculating, determine which student, Fiona or Gregor, accumulated 100 beans first. Explain how you know.

    b. Using Fiona's model ...

        i. How many trials would be needed for her to accumulate 100 beans?

    ii.   How many trials would be needed for her to accumulate 1,000 beans?

c.   Using Gregor's model ...

    i.   How many trials would be needed for him to accumulate 100 beans?

    ii.   How many trials would be needed for him to accumulate 1,000 beans?

d.   Was your prediction in part (a) correct? If not, what was the error in your reasoning?

      Lesson 24:   Solving Exponential Equations

       EUREKA MATH

2. Fiona wants to know when her model $f(t) = 1.263(1.357)^t$ predicts accumulations of 500, 5,000, and 50,000 beans, but she wants to find a way to figure it out without doing the same calculation three times.

a. Let the positive number c represent the number of beans that Fiona wants to have. Then solve the equation $1.263(1.357)^t = c$ for t.

b. Your answer to part (a) can be written as a function M of the number of beans c, where $c > 0$. Explain what this function represents.

c. When does Fiona's model predict that she will accumulate ...

i. 500 beans?

ii. 5,000 beans?

iii. 50,000 beans?

EUREKA
MATH™

Lesson 24:   Solving Exponential Equations

S.179

©2015 Great Minds. eureka-math.org
ALG II-M3-SE-B2-1.3.0-08.2015

3. Gregor states that the function g that he found to model his bean-flipping data can be written in the form $g(t) = 0.972\left(10^{\log(1.629)t}\right)$. Since $\log(1.629) \approx 0.2119$, he is using $g(t) = 0.972(10^{0.2119t})$ as his new model.

   a. Is Gregor correct? Is $g(t) = 0.972\left(10^{\log(1.629)t}\right)$ an equivalent form of his original function? Use properties of exponents and logarithms to explain how you know.

   b. Gregor also wants to find a function to help him to calculate the number of trials his function g predicts it should take to accumulate 500, 5,000, and 50,000 beans. Let the positive number c represent the number of beans that Gregor wants to have. Solve the equation $0.972(10^{0.2119t}) = c$ for t.

   c. Your answer to part (b) can be written as a function N of the number of beans c, where $c > 0$. Explain what this function represents.

   d. When does Gregor's model predict that he will accumulate ...
      i. 500 beans?

Lesson 24:   Solving Exponential Equations

©2015 Great Minds. eureka-math.org
ALG II-M3-SE-B2-1.3.0-08.2015

  ii. 5,000 beans?

  iii. 50,000 beans?

4. Helena and Karl each change the rules for the bean experiment. Helena started with four beans in her cup and added one bean for each that landed marked-side up for each trial. Karl started with one bean in his cup but added two beans for each that landed marked-side up for each trial.

  a. Helena modeled her data by the function $h(t) = 4.127(1.468^t)$. Explain why her values of $a = 4.127$ and $b = 1.468$ are reasonable.

  b. Karl modeled his data by the function $k(t) = 0.897(1.992^t)$. Explain why his values of $a = 0.897$ and $b = 1.992$ are reasonable.

c. At what value of t do Karl and Helena have the same number of beans?

d. Use a graphing utility to graph $y = h(t)$ and $y = k(t)$ for $0 < t < 10$.

e. Explain the meaning of the intersection point of the two curves $y = h(t)$ and $y = k(t)$ in the context of this problem.

f. Which student reaches 20 beans first? Does the reasoning used in deciding whether Gregor or Fiona would get 100 beans first hold true here? Why or why not?

Lesson 24:   Solving Exponential Equations

EUREKA
MATH

For the following functions f and g, solve the equation $f(x) = g(x)$. Express your solutions in terms of logarithms.

5. $f(x) = 10(3.7)^{x+1}$, $g(x) = 5(7.4)^x$

6. $f(x) = 135(5)^{3x+1}$, $g(x) = 75(3)^{4-3x}$

7. $f(x) = 100^{x^3+x^2-4x}$, $g(x) = 10^{2x^2-6x}$

8. $f(x) = 48\left(4^{x^2+3x}\right)$, $g(x) = 3\left(8^{x^2+4x+4}\right)$

9. $f(x) = e^{\sin^2(x)}$, $g(x) = e^{\cos^2(x)}$

10. $f(x) = (0.49)^{\cos(x)+\sin(x)}$, $g(x) = (0.7)^{2\sin(x)}$

EUREKA
MATH™

©2015 Great Minds. eureka-math.org
ALG II-M3-SE-B2-1.3.0-08.2015

Problem Set

1. Solve the following equations.

   a. $2 \cdot 5^{x+3} = 6250$

   b. $3 \cdot 6^{2x} = 648$

   c. $5 \cdot 2^{3x+5} = 10240$

   d. $4^{3x-1} = 32$

   e. $3 \cdot 2^{5x} = 216$

   f. $5 \cdot 11^{3x} = 120$

   g. $7 \cdot 9^{x} = 5405$

   h. $\sqrt{3} \cdot 3^{3x} = 9$

   i. $\log(400) \cdot 8^{5x} = \log(160000)$

2. Lucy came up with the model $f(t) = 0.701(1.382)^{t}$ for the first bean activity. When does her model predict that she would have 1,000 beans?

3. Jack came up with the model $g(t) = 1.033(1.707)^{t}$ for the first bean activity. When does his model predict that he would have 50,000 beans?

4. If instead of beans in the first bean activity you were using fair pennies, when would you expect to have $1,000,000?

5. Let $f(x) = 2 \cdot 3^{x}$ and $g(x) = 3 \cdot 2^{x}$.

   a. Which function is growing faster as x increases? Why?

   b. When will $f(x) = g(x)$?

6. The growth of a population of *E. coli* bacteria can be modeled by the function $E(t) = 500(11.547)^{t}$, and the growth of a population of *Salmonella* bacteria can be modeled by the function $S(t) = 4000(3.668)^{t}$, where t measures time in hours.

   a. Graph these two functions on the same set of axes. At which value of t does it appear that the graphs intersect?

   b. Use properties of logarithms to find the time t when these two populations are the same size. Give your answer to two decimal places.

7. Chain emails contain a message suggesting you will have bad luck if you do not forward the email to others. Suppose a student started a chain email by sending the message to 10 friends and asking those friends to each send the same email to 3 more friends exactly one day after receiving the message. Assuming that everyone that gets the email participates in the chain, we can model the number of people who receive the email on the nth day by the formula $E(n) = 10(3^n)$, where $n = 0$ indicates the day the original email was sent.

   a. If we assume the population of the United States is 318 million people and everyone who receives the email sends it to 3 people who have not received it previously, how many days until there are as many emails being sent out as there are people in the United States?

   b. The population of Earth is approximately 7.1 billion people. On what day will 7.1 billion emails be sent out?

8. Solve the following exponential equations.

   a. $10^{(3x-5)} = 7^x$

   b. $3^{\frac{x}{5}} = 2^{4x-2}$

   c. $10^{x^2+5} = 100^{2x^2+x+2}$

   d. $4^{x^2-3x+4} = 2^{5x-4}$

9. Solve the following exponential equations.

   a. $(2^x)^x = 8^x$

   b. $(3^x)^x = 12$

10. Solve the following exponential equations.

   a. $10^{x+1} - 10^{x-1} = 1287$

   b. $2(4^x) + 4^{x+1} = 342$

11. Solve the following exponential equations.

   a. $(10^x)^2 - 3(10^x) + 2 = 0$ Hint: Let $u = 10^x$, and solve for u before solving for x.

   b. $(2^x)^2 - 3(2^x) - 4 = 0$

   c. $3(e^x)^2 - 8(e^x) - 3 = 0$

   d. $4^x + 7(2^x) + 12 = 0$

   e. $(10^x)^2 - 2(10^x) - 1 = 0$

12. Solve the following systems of equations.

   a. $2^{x+2y} = 8$
      $4^{2x+y} = 1$

   b. $2^{2x+y-1} = 32$
      $4^{x-2y} = 2$

   c. $2^{3x} = 8^{2y+1}$
      $9^{2y} = 3^{3x-9}$

13. Because $f(x) = \log_b(x)$ is an increasing function, we know that if $p < q$, then $\log_b(p) < \log_b(q)$. Thus, if we take logarithms of both sides of an inequality, then the inequality is preserved. Use this property to solve the following inequalities.

   a. $4^x > \dfrac{5}{3}$

   b. $\left(\dfrac{2}{7}\right)^x > 9$

   c. $4^x > 8^{x-1}$

   d. $3^{x+2} > 5^{3-2x}$

   e. $\left(\dfrac{3}{4}\right)^x > \left(\dfrac{4}{3}\right)^{x+1}$

EUREKA
MATH

Lesson 24:   Solving Exponential Equations

S.187

©2015 Great Minds. eureka-math.org
ALG II-M3-SE-B2-1.3.0-08.2015

This page intentionally left blank

Lesson 25: Geometric Sequences and Exponential Growth and Decay

Classwork

Opening Exercise

Suppose a ball is dropped from an initial height h_0 and that each time it rebounds, its new height is 60% of its previous height.

a. What are the first four rebound heights h_1, h_2, h_3, and h_4 after being dropped from a height of $h_0 = 10$ ft.?

b. Suppose the initial height is A ft. What are the first four rebound heights? Fill in the following table:

Rebound	Height (ft.)
1	
2	
3	
4	

c. How is each term in the sequence related to the one that came before it?

d. Suppose the initial height is A ft. and that each rebound, rather than being 60% of the previous height, is r times the previous height, where $0 < r < 1$. What are the first four rebound heights? What is the nth rebound height?

EUREKA
MATH™

©2015 Great Minds. eureka-math.org
ALG II-M3-SE-B2-1.3.0-08.2015

e.   What kind of sequence is the sequence of rebound heights?

f.   Suppose that we define a function f with domain the positive integers so that $f(1)$ is the first rebound height, $f(2)$ is the second rebound height, and continuing so that $f(k)$ is the kth rebound height for positive integers k. What type of function would you expect f to be?

g.   On the coordinate plane below, sketch the height of the bouncing ball when $A = 10$ and $r = 0.60$, assuming that the highest points occur at $x = 1, 2, 3, 4, \ldots$.

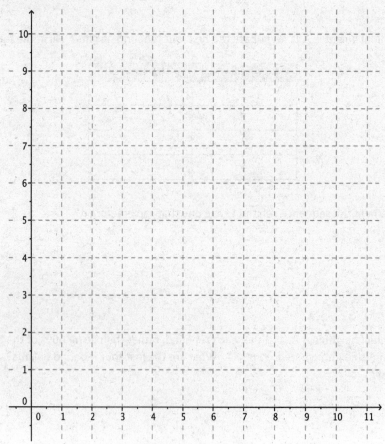

Lesson 25:   Geometric Sequences and Exponential Growth and Decay

EUREKA
MATH

h. Does the exponential function $f(x) = 10(0.60)^x$ for real numbers x model the height of the bouncing ball? Explain how you know.

i. What does the function $f(n) = 10(0.60)^n$ for integers $n \geq 0$ model?

Exercises

1. Jane works for a videogame development company that pays her a starting salary of $100 per day, and each day she works she earns $100 more than the day before.

    a. How much does she earn on day 5?

    b. If you were to graph the growth of her salary for the first 10 days she worked, what would the graph look like?

    c. What kind of sequence is the sequence of Jane's earnings each day?

2. A laboratory culture begins with 1,000 bacteria at the beginning of the experiment, which we denote by time 0 hours. By time 2 hours, there are 2,890 bacteria.

   a. If the number of bacteria is increasing by a common factor each hour, how many bacteria are there at time 1 hour? At time 3 hours?

   b. Find the explicit formula for term P_n of the sequence in this case.

   c. How would you find term P_{n+1} if you know term P_n? Write a recursive formula for P_{n+1} in terms of P_n.

   d. If P_0 is the initial population, the growth of the population P_n at time n hours can be modeled by the sequence $P_n = P(n)$, where P is an exponential function with the following form:

   $$P(n) = P_0 2^{kn}, \text{ where } k > 0.$$

   Find the value of k and write the function P in this form. Approximate k to four decimal places.

   e. Use the function in part (d) to determine the value of t when the population of bacteria has doubled.

EUREKA
MATH™

f. If P_0 is the initial population, the growth of the population P at time t can be expressed in the following form:

$$P(n) = P_0 e^{kn}, \text{ where } k > 0.$$

Find the value of k, and write the function P in this form. Approximate k to four decimal places.

g. Use the formula in part (d) to determine the value of t when the population of bacteria has doubled.

3. The first term a_0 of a geometric sequence is -5, and the common ratio r is -2.

a. What are the terms a_0, a_1, and a_2?

b. Find a recursive formula for this sequence.

c. Find an explicit formula for this sequence.

d. What is term a_9?

e. What is term a_{10}?

4. Term a_4 of a geometric sequence is 5.8564, and term a_5 is -6.44204.

   a. What is the common ratio r?

   b. What is term a_0?

   c. Find a recursive formula for this sequence.

   d. Find an explicit formula for this sequence.

5. The recursive formula for a geometric sequence is $a_{n+1} = 3.92(a_n)$ with $a_0 = 4.05$. Find an explicit formula for this sequence.

6. The explicit formula for a geometric sequence is $a_n = 147(2.1)^{3n}$. Find a recursive formula for this sequence.

EUREKA
MATH™

©2015 Great Minds. eureka-math.org
ALG II-M3-SE-B2-1.3.0-08.2015

Lesson Summary

ARITHMETIC SEQUENCE: A sequence is called *arithmetic* if there is a real number d such that each term in the sequence is the sum of the previous term and d.

- *Explicit formula:* Term a_n of an arithmetic sequence with first term a_0 and common difference d is given by $a_n = a_0 + nd$, for $n \geq 0$.

- *Recursive formula:* Term a_{n+1} of an arithmetic sequence with first term a_0 and common difference d is given by $a_{n+1} = a_n + d$, for $n \geq 0$.

GEOMETRIC SEQUENCE: A sequence is called *geometric* if there is a real number r such that each term in the sequence is a product of the previous term and r.

- *Explicit formula:* Term a_n of a geometric sequence with first term a_0 and common ratio r is given by $a_n = a_0 r^n$, for $n \geq 0$.

- *Recursive formula:* Term a_{n+1} of a geometric sequence with first term a_0 and common ratio r is given by $a_{n+1} = a_n r$.

Problem Set

1. Convert the following recursive formulas for sequences to explicit formulas.

   a. $a_{n+1} = 4.2 + a_n$ with $a_0 = 12$

   b. $a_{n+1} = 4.2 a_n$ with $a_0 = 12$

   c. $a_{n+1} = \sqrt{5}\, a_n$ with $a_0 = 2$

   d. $a_{n+1} = \sqrt{5} + a_n$ with $a_0 = 2$

   e. $a_{n+1} = \pi\, a_n$ with $a_0 = \pi$

2. Convert the following explicit formulas for sequences to recursive formulas.

   a. $a_n = \frac{1}{5}(3^n)$ for $n \geq 0$

   b. $a_n = 16 - 2n$ for $n \geq 0$

   c. $a_n = 16\left(\frac{1}{2}\right)^n$ for $n \geq 0$

   d. $a_n = 71 - \frac{6}{7}n$ for $n \geq 0$

   e. $a_n = 190(1.03)^n$ for $n \geq 0$

3. If a geometric sequence has $a_1 = 256$ and $a_8 = 512$, find the exact value of the common ratio r.

4. If a geometric sequence has $a_2 = 495$ and $a_6 = 311$, approximate the value of the common ratio r to four decimal places.

5. Find the difference between the terms a_{10} of an arithmetic sequence and a geometric sequence, both of which begin at term a_0 and have $a_2 = 4$ and $a_4 = 12$.

6. Given the geometric sequence defined by the following values of a_0 and r, find the value of n so that a_n has the specified value.

   a. $a_0 = 64, r = \frac{1}{2}, a_n = 2$

   b. $a_0 = 13, r = 3, a_n = 85293$

   c. $a_0 = 6.7, r = 1.9, a_n = 7804.8$

   d. $a_0 = 10958, r = 0.7, a_n = 25.5$

7. Jenny planted a sunflower seedling that started out 5 cm tall, and she finds that the average daily growth is 3.5 cm.

   a. Find a recursive formula for the height of the sunflower plant on day n.

   b. Find an explicit formula for the height of the sunflower plant on day $n \geq 0$.

8. Kevin modeled the height of his son (in inches) at age n years for $n = 2, 3, \ldots, 8$ by the sequence $h_n = 34 + 3.2(n - 2)$. Interpret the meaning of the constants 34 and 3.2 in his model.

9. Astrid sells art prints through an online retailer. She charges a flat rate per order for an order processing fee, sales tax, and the same price for each print. The formula for the cost of buying n prints is given by $P_n = 4.5 + 12.6n$.

   a. Interpret the number 4.5 in the context of this problem.

   b. Interpret the number 12.6 in the context of this problem.

   c. Find a recursive formula for the cost of buying n prints.

10. A bouncy ball rebounds to 90% of the height of the preceding bounce. Craig drops a bouncy ball from a height of 20 feet.

    a. Write out the sequence of the heights h_1, h_2, h_3, and h_4 of the first four bounces, counting the initial height as $h_0 = 20$.

    b. Write a recursive formula for the rebound height of a bouncy ball dropped from an initial height of 20 feet.

    c. Write an explicit formula for the rebound height of a bouncy ball dropped from an initial height of 20 feet.

    d. How many bounces will it take until the rebound height is under 6 feet?

    e. Extension: Find a formula for the minimum number of bounces needed for the rebound height to be under y feet, for a real number $0 < y < 20$.

11. Show that when a quantity $a_0 = A$ is increased by $x\%$, its new value is $a_1 = A\left(1 + \frac{x}{100}\right)$. If this quantity is again increased by $x\%$, what is its new value a_2? If the operation is performed n times in succession, what is the final value of the quantity a_n?

Lesson 25:   Geometric Sequences and Exponential Growth and Decay

12. When Eli and Daisy arrive at their cabin in the woods in the middle of winter, the interior temperature is 40°F.

   a.   Eli wants to turn up the thermostat by 2°F every 15 minutes. Find an explicit formula for the sequence that represents the thermostat settings using Eli's plan.

   b.   Daisy wants to turn up the thermostat by 4% every 15 minutes. Find an explicit formula for the sequence that represents the thermostat settings using Daisy's plan.

   c.   Which plan gets the thermostat to 60°F most quickly?

   d.   Which plan gets the thermostat to 72°F most quickly?

13. In nuclear fission, one neutron splits an atom causing the release of two other neutrons, each of which splits an atom and produces the release of two more neutrons, and so on.

   a.   Write the first few terms of the sequence showing the numbers of atoms being split at each stage after a single atom splits. Use $a_0 = 1$.

   b.   Find the explicit formula that represents your sequence in part (a).

   c.   If the interval from one stage to the next is one-millionth of a second, write an expression for the number of atoms being split at the end of one second.

   d.   If the number from part (c) were written out, how many digits would it have?

This page intentionally left blank

Lesson 26: Percent Rate of Change

Classwork

Exercise

Answer the following questions.

The youth group from Example 1 is given the option of investing their money at 2.976% interest per year, compounded monthly instead of depositing it in the original account earning 3.0% compounded yearly.

a. With an initial deposit of $800, how much would be in each account after two years?

b. Compare the total amount from part (a) to how much they would have made using the interest rate of 3% compounded yearly for two years. Which account would you recommend the youth group invest its money in? Why?

Lesson Summary

- For application problems involving a percent rate of change represented by the unit rate r, we can write $F(t) = P(1 + r)^t$, where F is the future value (or ending amount), P is the present amount, and t is the number of time units. When the percent rate of change is negative, r is negative and the quantity decreases with time.

- The nominal APR is the percent rate of change per compounding period times the number of compounding periods per year. If the nominal APR is given by the unit rate r and is compounded n times a year, then function $F(t) = P\left(1 + \frac{r}{n}\right)^{nt}$ describes the future value at time t of an account given that nominal APR and an initial value of P.

- For continuous compounding, we can write $F = Pe^{rt}$, where e is Euler's number and r is the unit rate associated to the percent rate of change.

Problem Set

1. Write each recursive sequence in explicit form. Identify each sequence as arithmetic, geometric, or neither.

   a. $a_1 = 3, a_{n+1} = a_n + 5$

   b. $a_1 = -1, a_{n+1} = -2a_n$

   c. $a_1 = 30, a_{n+1} = a_n - 3$

   d. $a_1 = \sqrt{2}, a_{n+1} = \frac{a_n}{\sqrt{2}}$

   e. $a_1 = 1, a_{n+1} = \cos(\pi a_n)$

2. Write each sequence in recursive form. Assume the first term is when $n = 1$.

   a. $a_n = \frac{3}{2}n + 3$

   b. $a_n = 3\left(\frac{3}{2}\right)^n$

   c. $a_n = n^2$

   d. $a_n = \cos(2\pi n)$

3. Consider two bank accounts. Bank A gives simple interest on an initial investment in savings accounts at a rate of 3% per year. Bank B gives compound interest on savings accounts at a rate of 2.5% per year. Fill out the following table.

Number of Years, n	Bank A Balance, a_n (in dollars)	Bank B Balance, b_n (in dollars)
0	1,000.00	1,000.00
1		
2		
3		
4		
5		

a. What type of sequence do the Bank A balances represent?

b. Give both a recursive and an explicit formula for the Bank A balances.

c. What type of sequence do the Bank B balances represent?

d. Give both a recursive and an explicit formula for the Bank B balances.

e. Which bank account balance is increasing faster in the first five years?

f. If you were to recommend a bank account for a long-term investment, which would you recommend?

g. At what point is the balance in Bank B larger than the balance in Bank A?

4. You decide to invest your money in a bank that uses continuous compounding at 5.5% interest per year. You have $500.

a. Ja'mie decides to invest $1,000 in the same bank for one year. She predicts she will have double the amount in her account than you will have. Is this prediction correct? Explain.

b. Jonas decides to invest $500 in the same bank as well, but for two years. He predicts that after two years he will have double the amount of cash that you will after one year. Is this prediction correct? Explain.

5. Use the properties of exponents to identify the percent rate of change of the functions below, and classify them as representing exponential growth or decay. (The first two problems are done for you.)

a. $f(t) = (1.02)^t$

b. $f(t) = (1.01)^{12t}$

c. $f(t) = (0.97)^t$

d. $f(t) = 1000(1.2)^t$

e. $f(t) = \dfrac{(1.07)^t}{1000}$

f. $f(t) = 100 \cdot 3^t$

g. $f(t) = 1.05 \cdot \left(\dfrac{1}{2}\right)^t$

h. $f(t) = 80 \cdot \left(\dfrac{49}{64}\right)^{\frac{1}{2}t}$

i. $f(t) = 1.02 \cdot (1.13)^{\pi t}$

6. The effective rate of an investment is the percent rate of change per year associated to the nominal APR. The effective rate is very useful in comparing accounts with different interest rates and compounding periods. In general, the effective rate can be found with the following formula: $r_E = \left(1 + \frac{r}{k}\right)^k - 1$. The effective rate presented here is the interest rate needed for annual compounding to be equal to compounding n times per year.

    a. For investing, which account is better: an account earning a nominal APR of 7% compounded monthly or an account earning a nominal APR of 6.875% compounded daily? Why?

    b. The effective rate formula for an account compounded continuously is $r_E = e^r - 1$. Would an account earning 6.875% interest compounded continuously be better than the accounts in part (a)?

7. Radioactive decay is the process in which radioactive elements decay into more stable elements. A half-life is the time it takes for half of an amount of an element to decay into a more stable element. For instance, the half-life of uranium-235 is 704 million years. This means that half of any sample of uranium-235 will transform into lead-207 in 704 million years. We can assume that radioactive decay is modeled by exponential decay with a constant decay rate.

    a. Suppose we have a sample of A g of uranium-235. Write an exponential formula that gives the amount of uranium-235 remaining after m half-lives.

    b. Does the formula that you wrote in part (a) work for any radioactive element? Why?

    c. Suppose we have a sample of A g of uranium-235. What is the decay rate per million years? Write an exponential formula that gives the amount of uranium-235 remaining after t million years.

    d. How would you calculate the number of years it takes to get to a specific percentage of the original amount of material? For example, how many years will it take for there to be 80% of the original amount of uranium-235 remaining?

    e. How many millions of years would it take 2.35 kg of uranium-235 to decay to 1 kg of uranium?

8. Doug drank a cup of tea with 130 mg of caffeine. Each hour, the caffeine in Doug's body diminishes by about 12%. (This rate varies between 6% and 14% depending on the person.)

    a. Write a formula to model the amount of caffeine remaining in Doug's system after each hour.

    b. About how long should it take for the level of caffeine in Doug's system to drop below 30 mg?

    c. The time it takes for the body to metabolize half of a substance is called its half-life. To the nearest 5 minutes, how long is the half-life for Doug to metabolize caffeine?

    d. Write a formula to model the amount of caffeine remaining in Doug's system after m half-lives.

9. A study done from 1950 through 2000 estimated that the world population increased on average by 1.77% each year. In 1950, the world population was 2.519 billion.

    a. Write a function p for the world population t years after 1950.

    b. If this trend continued, when should the world population have reached 7 billion?

    c. The world population reached 7 billion October 31, 2011, according to the United Nations. Is the model reasonably accurate?

    d. According to the model, when should the world population be greater than 12 billion people?

©2015 Great Minds. eureka-math.org
ALG II-M3-SE-B2-1.3.0-08.2015

10. A particular mutual fund offers 4.5% nominal APR compounded monthly. Trevor wishes to deposit $1,000.

   a. What is the percent rate of change per month for this account?

   b. Write a formula for the amount Trevor will have in the account after m months.

   c. *Doubling time* is the amount of time it takes for an investment to double. What is the doubling time of Trevor's investment?

11. When paying off loans, the monthly payment first goes to any interest owed before being applied to the remaining balance. Accountants and bankers use tables to help organize their work.

   a. Consider the situation that Fred is paying off a loan of $125,000 with an interest rate of 6% per year compounded monthly. Fred pays $749.44 every month. Complete the following table:

Payment	Interest Paid	Principal Paid	Remaining Principal
$749.44			
$749.44			
$749.44			

   b. Fred's loan is supposed to last for 30 years. How much will Fred end up paying if he pays $749.44 every month for 30 years? How much of this is interest if his loan was originally for $125,000?

This page intentionally left blank

Lesson 27: Modeling with Exponential Functions

Classwork

Opening Exercise

The following table contains U.S. population data for the two most recent census years, 2000 and 2010.

Census Year	U.S. Population (in millions)
2000	281.4
2010	308.7

a. Steve thinks the data should be modeled by a linear function.

    i. What is the average rate of change in population per year according to this data?

    ii. Write a formula for a linear function, L, to estimate the population t years since the year 2000.

b. Phillip thinks the data should be modeled by an exponential function.

    i. What is the growth rate of the population per year according to this data?

    ii. Write a formula for an exponential function, E, to estimate the population t years since the year 2000.

c.   Who has the correct model? How do you know?

Mathematical Modeling Exercises 1–14

This challenge continues to examine U.S. census data to select and refine a model for the population of the United States over time.

1.   The following table contains additional U.S. census population data. Would it be more appropriate to model this data with a linear or an exponential function? Explain your reasoning.

Census Year	U.S. Population (in millions of people)
1900	76.2
1910	92.2
1920	106.0
1930	122.8
1940	132.2
1950	150.7
1960	179.3
1970	203.3
1980	226.5
1990	248.7
2000	281.4
2010	308.7

2.   Use a calculator's regression capability to find a function, f, that models the US Census Bureau data from 1900 to 2010.

©2015 Great Minds. eureka-math.org
ALG II-M3-SE-B2-1.3.0-08.2015

3. Find the growth factor for each 10-year period and record it in the table below. What do you observe about these growth factors?

Census Year	U.S. Population (in millions of people)	Growth Factor (10-year period)
1900	76.2	--
1910	92.2	
1920	106.0	
1930	122.8	
1940	132.2	
1950	150.7	
1960	179.3	
1970	203.3	
1980	226.5	
1990	248.7	
2000	281.4	
2010	308.7	

4. For which decade is the 10-year growth factor the lowest? What factors do you think caused that decrease?

5. Find an average 10-year growth factor for the population data in the table. What does that number represent? Use the average growth factor to find an exponential function, g, that can model this data.

6. You have now computed three potential models for the population of the United States over time: functions E, f, and g. Which one do you expect would be the most accurate model based on how they were created? Explain your reasoning.

©2015 Great Minds. eureka-math.org
ALG II-M3-SE-B2-1.3.0-08.2015

7. Summarize the three formulas for exponential models that you have found so far: Write the formula, the initial populations, and the growth rates indicated by each function. What is different between the structures of these three functions?

8. Rewrite the functions E, f, and g as needed in terms of an annual growth rate.

9. Transform the functions as needed so that the time $t = 0$ represents the same year in functions E, f, and g. Then compare the values of the initial populations and annual growth rates indicated by each function.

10. Which of the three functions is the best model to use for the U.S. census data from 1900 to 2010? Explain your reasoning.

11. The U.S. Census Bureau website http://www.census.gov/popclock displays the current estimate of both the United States and world populations.

    a. What is today's current estimated population of the U.S.?

    b. If time $t = 0$ represents the year 1900, what is the value of t for today's date? Give your answer to two decimal places.

    c. Which of the functions E, f, and g gives the best estimate of today's population? Does that match what you expected? Justify your reasoning.

    d. With your group, discuss some possible reasons for the discrepancy between what you expected in Exercise 8 and the results of part (c) above.

12. Use the model that most accurately predicted today's population in Exercise 9, part (c) to predict when the U.S. population will reach half a billion.

13. Based on your work so far, do you think this is an accurate prediction? Justify your reasoning.

14. Here is a graph of the U.S. population since the census began in 1790. Which type of function would best model this data? Explain your reasoning.

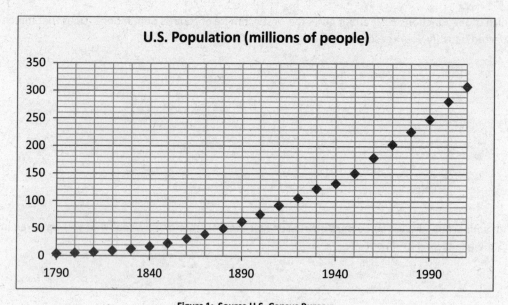

Figure 1: Source U.S. Census Bureau

Exercises 15–16

15. The graph below shows the population of New York City during a time of rapid population growth.

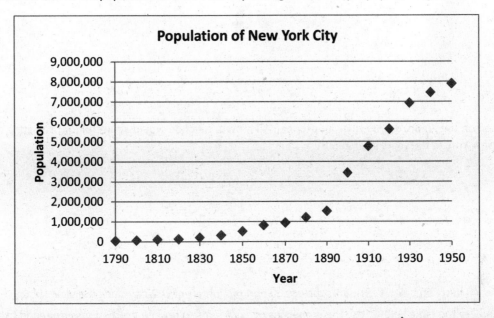

Finn averaged the 10-year growth rates and wrote the function $f(t) = 33\,31(1.44)^{\frac{t}{10}}$, where t is the time in years since 1790.

Gwen used the regression features on a graphing calculator and got the function $g(t) = 48661(1.036)^t$, where t is the time in years since 1790.

a. Rewrite each function to determine the annual growth rate for Finn's model and Gwen's model.

b. What is the predicted population in the year 1790 for each model?

c. Lenny calculated an exponential regression using his graphing calculator and got the same growth rate as Gwen, but his initial population was very close to 0. Explain what data Lenny may have used to find his function.

EUREKA
MATH™

©2015 Great Minds. eureka-math.org
ALG II-M3-SE-B2-1.3.0-08.2015

d. When does Gwen's function predict the population will reach 1,000 ,000? How does this compare to the graph?

e. Based on the graph, do you think an exponential growth function would be useful for predicting the population of New York in the years after 1950?

16. Suppose each function below represents the population of a different U.S. city since the year 1900.

a. Complete the table below. Use the properties of exponents to rewrite expressions as needed to help support your answers.

City Population Function (t is years since 1900)	Population in the Year 1900	Annual Growth/Decay Rate	Predicted in 2000	Between Which Years Did the Population Double?
$A(t) = 3000(1.1)^{\frac{t}{5}}$				
$B(t) = \dfrac{(1.5)^{2t}}{2.25}$				
$C(t) = 10000(1 - 0.01)^t$				
$D(t) = 900(1.02)^t$				

Lesson 27:   Modeling with Exponential Functions

©2015 Great Minds. eureka-math.org
ALG II-M3-SE-B2-1.3.0-08.2015

b. Could the function $E(t) = 6520(1.219)^{\frac{t}{10}}$, where t is years since 2000 also represent the population of one of these cities? Use the properties of exponents to support your answer.

c. Which cities are growing in size and which are decreasing according to these models?

d. Which of these functions might realistically represent city population growth over an extended period of time?

Lesson Summary

To model exponential data as a function of time:

- Examine the data to see if there appears to be a constant growth or decay factor.
- Determine a growth factor and a point in time to correspond to $t = 0$.

- Create a function $f(t) = a \cdot b^{ct}$ to model the situation, where b is the growth factor every $\frac{1}{c}$ years and a is the value of f when $t = 0$.

Logarithms can be used to solve for t when you know the value of $f(t)$ in an exponential function.

Problem Set

1. Does each pair of formulas described below represent the same sequence? Justify your reasoning.

   a. $a_{n+1} = \frac{2}{3}a_n$, $a_0 = -1$ and $b_n = -\left(\frac{2}{3}\right)^n$ for $n \geq 0$.

   b. $a_n = 2a_{n-1} + 3$, $a_0 = 3$ and $b_n = 2(n-1)^3 + 4(n-1) + 3$ for $n \geq 1$.

   c. $a_n = \frac{1}{3}(3)^n$ for $n \geq 0$ and $b_n = 3^{n-2}$ for $n \geq 0$.

2. Tina is saving her babysitting money. She has \$500 in the bank, and each month she deposits another \$100. Her account earns 2% interest compounded monthly.

   a. Complete the table showing how much money she has in the bank for the first four months.

Month	Amount (in dollars)
1	
2	
3	
4	

   b. Write a recursive sequence for the amount of money she has in her account after n months.

---

3. Assume each table represents values of an exponential function of the form $f(t) = a(b)^{ct}$, where b is a positive real number and a and c are real numbers. Use the information in each table to write a formula for f in terms of t for parts (a)–(d).

a.
t	$f(t)$
0	10
4	50

b.
t	$f(t)$
0	1000
5	750

c.
t	$f(t)$
6	25
8	45

d.
t	$f(t)$
3	50
6	40

e. Rewrite the expressions for each function in parts (a)–(d) to determine the annual growth or decay rate.

f. For parts (a) and (c), determine when the value of the function is double its initial amount.

g. For parts (b) and (d), determine when the value of the function is half of its initial amount.

4. When examining the data in Example 1, Juan noticed the population doubled every five years and wrote the formula $P(t) = 100(2)^{\frac{t}{5}}$. Use the properties of exponents to show that both functions grow at the same rate per year.

5. The growth of a tree seedling over a short period of time can be modeled by an exponential function. Suppose the tree starts out 3 feet tall and its height increases by 15% per year. When will the tree be 25 feet tall?

6. Loggerhead turtles reproduce every 2–4 years, laying approximately 120 eggs in a clutch. Studying the local population, a biologist records the following data in the second and fourth years of her study:

Year	Population
2	50
4	1250

a. Find an exponential model that describes the loggerhead turtle population in year t.

b. According to your model, when will the population of loggerhead turtles be over 5,000? Give your answer in years and months.

7. The radioactive isotope seaborgium-266 has a half-life of 30 seconds, which means that if you have a sample of A grams of seaborgium-266, then after 30 seconds half of the sample has decayed (meaning it has turned into another element) and only $\frac{A}{2}$ grams of seaborgium-266 remain. This decay happens continuously.

a. Define a sequence $a_0, a_1, a_2, \ldots$ so that a_n represents the amount of a 100-gram sample that remains after n minutes.

b. Define a function $a(t)$ that describes the amount of a 100-gram sample of seaborgium-266 that remains after t minutes.

c. Do your sequence from part (a) and your function from part (b) model the same thing? Explain how you know.

d. How many minutes does it take for less than 1 g of seaborgium-266 to remain from the original 100 g sample? Give your answer to the nearest minute.

8.   Strontium-90, magnesium-28, and bismuth all decay radioactively at different rates. Use data provided in the graphs and tables below to answer the questions that follow. .

Strontium-90 (grams) vs. time (hours)

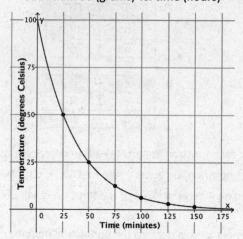

Radioactive Decay of Magnesium-28	
R grams	t hours
1	0
0.5	21
0.25	42
0.125	63
0.0625	84

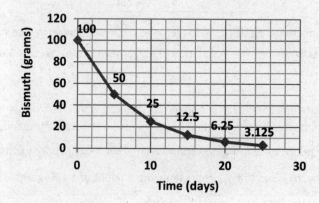

a.   Which element decays most rapidly? How do you know?

b.   Write an exponential function for each element that shows how much of a 100 g sample will remain after t days. Rewrite each expression to show precisely how their exponential decay rates compare to confirm your answer to part (a).

9. The growth of two different species of fish in a lake can be modeled by the functions shown below where t is time in months since January 2000. Assume these models will be valid for at least 5 years.

    Fish A: $f(t) = 5000(1.3)^t$

    Fish B: $g(t) = 10000(1.1)^t$

    According to these models, explain why the fish population modeled by function f will eventually catch up to the fish population modeled by function g. Determine precisely when this will occur.

10. When looking at U.S. minimum wage data, you can consider the nominal minimum wage, which is the amount paid in dollars for an hour of work in the given year. You can also consider the minimum wage adjusted for inflation. Below is a table showing the nominal minimum wage and a graph of the data when the minimum wage is adjusted for inflation. Do you think an exponential function would be an appropriate model for either situation? Explain your reasoning.

Year	Nominal Minimum Wage
1940	$0.30
1945	$0.40
1950	$0.75
1955	$0.75
1960	$1.00
1965	$1.25
1970	$1.60
1975	$2.10
1980	$3.10
1985	$3.35
1990	$3.80
1995	$4.25
2000	$5.15
2005	$5.15
2010	$7.25

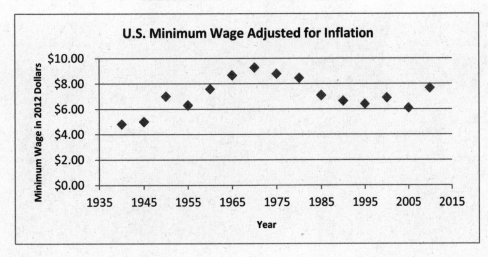

11. A dangerous bacterial compound forms in a closed environment but is immediately detected. An initial detection reading suggests the concentration of bacteria in the closed environment is one percent of the fatal exposure level. Two hours later, the concentration has increased to four percent of the fatal exposure level.

a. Develop an exponential model that gives the percentage of fatal exposure level in terms of the number of hours passed.

b. Doctors and toxicology professionals estimate that exposure to two-thirds of the bacteria's fatal concentration level will begin to cause sickness. Offer a time limit (to the nearest minute) for the inhabitants of the infected environment to evacuate in order to avoid sickness.

c. A prudent and more conservative approach is to evacuate the infected environment before bacteria concentration levels reach 45% of the fatal level. Offer a time limit (to the nearest minute) for evacuation in this circumstance.

d. To the nearest minute, when will the infected environment reach 100% of the fatal level of bacteria concentration?

12. Data for the number of users at two different social media companies is given below. Assuming an exponential growth rate, which company is adding users at a faster annual rate? Explain how you know.

Social Media Company A	
Year	Number of Users (Millions)
2010	54
2012	185

Social Media Company B	
Year	Number of Users (Millions)
2009	360
2012	1056

Lesson 28: Newton's Law of Cooling, Revisited

Classwork

Newton's law of cooling is used to model the temperature of an object placed in an environment of a different temperature. The temperature of the object t hours after being placed in the new environment is modeled by the formula

$$T(t) = T_a + (T_0 - T_a) \cdot e^{-kt},$$

where:

    $T(t)$ is the temperature of the object after a time of t hours has elapsed,

    T_a is the ambient temperature (the temperature of the surroundings), assumed to be constant and not impacted by the cooling process,

    T_0 is the initial temperature of the object, and

    k is the decay constant.

Mathematical Modeling Exercise 1

A crime scene investigator is called to the scene of a crime where a dead body has been found. He arrives at the scene and measures the temperature of the dead body at 9:30 p.m. to be 78.3°F. He checks the thermostat and determines that the temperature of the room has been kept at 74°F. At 10:30 p.m., the investigator measures the temperature of the body again. It is now 76.8°F. He assumes that the initial temperature of the body was 98.6°F (normal body temperature). Using this data, the crime scene investigator proceeds to calculate the time of death. According to the data he collected, what time did the person die?

    a.   Can we find the time of death using only the temperature measured at 9:30 p.m.? Explain.

    b.   Set up a system of two equations using the data.

d. Find the value of the decay constant, k.

e. What was the time of death?

Mathematical Modeling Exercise 2

A pot of tea is heated to 90°C. A cup of the tea is poured into a mug and taken outside where the temperature is 18°C. After 2 minutes, the temperature of the cup of tea is approximately 65°C.

a. Determine the value of the decay constant, k.

b. Write a function for the temperature of the tea in the mug, T, in °C, as a function of time, t, in minutes.

c.   Graph the function T.

d.   Use the graph of T to describe how the temperature decreases over time.

e. Use properties of exponents to rewrite the temperature function in the form $T(t) = 18 + 72(1 + r)^t$.

f. In Lesson 26, we saw that the value of r represents the percent change of a quantity that is changing according to an exponential function of the form $f(t) = A(1 + r)^t$. Describe what r represents in the context of the cooling tea.

g. As more time elapses, what temperature does the tea approach? Explain using both the context of the problem and the graph of the function T.

Mathematical Modeling Exercise 3

Two thermometers are sitting in a room that is 22°C. When each thermometer reads 22°C, the thermometers are placed in two different ovens. Data for the temperatures T_1 and T_2 of these thermometers (in °C) t minutes after being placed in the oven is provided below.

Thermometer 1:

t (minutes)	0	2	5	8	10	14
T_1 (°C)	22	75	132	173	175	176

Thermometer 2:

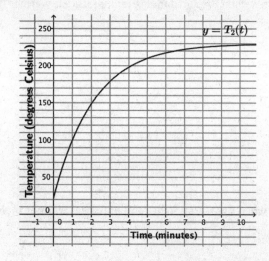

a. Do the table and graph given for each thermometer support the statement that Newton's law of cooling also applies when the surrounding temperature is warmer? Explain.

b. Which thermometer was placed in a hotter oven? Explain.

c. Using a generic decay constant, k, without finding its value, write an equation for each thermometer expressing the temperature as a function of time.

d. How do the equations differ when the surrounding temperature is warmer than the object rather than cooler as in previous examples?

e. How do the graphs differ when the surrounding temperature is warmer than the object rather than cooler as in previous examples?

Problem Set

1. Experiments with a covered cup of coffee show that the temperature (in degrees Fahrenheit) of the coffee can be modeled by the following equation:

$$f(t) = 112e^{-0.08t} + 68,$$

where the time is measured in minutes after the coffee was poured into the cup.

a. What is the temperature of the coffee at the beginning of the experiment?

b. What is the temperature of the room?

c. After how many minutes is the temperature of the coffee 140°F? Give your answer to 3 decimal places.

d. What is the temperature of the coffee after a few hours have elapsed?

e. What is the percent rate of change of the difference between the temperature of the room and the temperature of the coffee?

2. Suppose a frozen package of hamburger meat is removed from a freezer that is set at 0°F and placed in a refrigerator that is set at 38°F. Six hours after being placed in the refrigerator, the temperature of the meat is 12°F.

a. Determine the decay constant, k.

b. Write a function for the temperature of the meat, T in Fahrenheit, as a function of time, t in hours.

c. Graph the function T.

d. Describe the transformations required to graph the function T beginning with the graph of the natural exponential function $f(t) = e^t$.

e. How long will it take the meat to thaw (reach a temperature above 32°F)? Give answer to three decimal places.

f. What is the percent rate of change of the difference between the temperature of the refrigerator and the temperature of the meat?

3. The table below shows the temperature of a pot of soup that was removed from the stove at time $t = 0$.

t (min)	0	10	20	30	40	50	60
T (°C)	100	34.183	22.514	20.446	20.079	20.014	20.002

a. What is the initial temperature of the soup?

b. What does the ambient temperature (room temperature) appear to be?

c. Use the temperature at $t = 10$ minutes to find the decay constant, k.

d. Confirm the value of k by using another data point from the table.

e. Write a function for the temperature of the soup (in Celsius) as a function of time in minutes.

f. Graph the function T.

4. Match each verbal description with its correct graph and write a possible equation expressing temperature as a function of time.

    a. A pot of liquid is heated to a boil and then placed on a counter to cool.

    b. A frozen dinner is placed in a preheated oven to cook.

    c. A can of room-temperature soda is placed in a refrigerator.

(i)
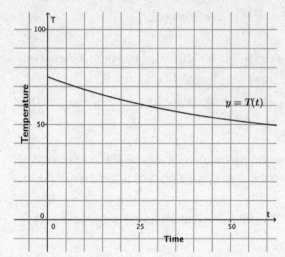

(ii)
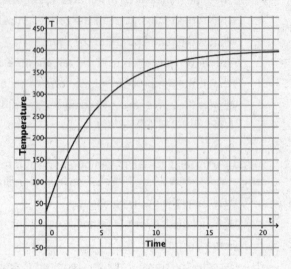

(iii)
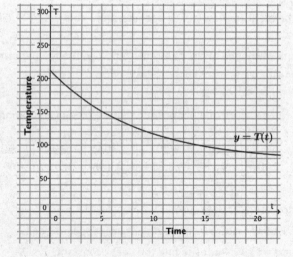

EUREKA
MATH™

Lesson 29: The Mathematics Behind a Structured Savings Plan

Classwork

Opening Exercise

Suppose you invested $1,000 in an account that paid an annual interest rate of 3% compounded monthly. How much would you have after 1 year?

Example 1

Let $a, ar, ar^2, ar^3, ar^4, \ldots$ be a geometric sequence with first term a and common ratio r. Show that the sum S_n of the first n terms of the geometric series

$$S_n = a + ar + ar^2 + ar^3 + \cdots + ar^{n-1} \quad (r \neq 1)$$

is given by the equation

$$S_n = a\left(\frac{1 - r^n}{1 - r}\right).$$

Exercises 1–3

1. Find the sum of the geometric series $3 + 6 + 12 + 24 + 48 + 96 + 192$.

2. Find the sum of the geometric series $40 + 40(1.005) + 40(1.005)^2 + \cdots + 40(1.005)^{11}$.

3. Describe a situation that might lead to calculating the sum of the geometric series in Exercise 2.

Example 2

A \$100 deposit is made at the end of every month for 12 months in an account that earns interest at an annual interest rate of 3% compounded monthly. How much will be in the account immediately after the last payment?

Discussion

An *annuity* is a series of payments made at fixed intervals of time. Examples of annuities include structured savings plans, lease payments, loans, and monthly home mortgage payments. The term annuity sounds like it is only a yearly payment, but annuities often require payments monthly, quarterly, or semiannually. The *future amount of the annuity*, denoted A_f, is the sum of all the individual payments made plus all the interest generated from those payments over the specified period of time.

We can generalize the structured savings plan example above to get a generic formula for calculating the future value of an annuity A_f in terms of the recurring payment R, interest rate i, and number of payment periods n. In the example above, we had a recurring payment of $R = 100$, an interest rate per time period of $i = 0.025$, and 12 payments, so $n = 12$. To make things simpler, we always assume that the payments and the time period in which interest is compounded are at the same time. That is, we do not consider plans where deposits are made halfway through the month with interest compounded at the end of the month.

In the example, the amount A_f of the structured savings plan annuity was the sum of all payments plus the interest accrued for each payment:

$$A_f = R + R(1 + i)^1 + R(1 + i)^2 + \cdots + R(1 + i)^{n-1}.$$

This, of course, is a geometric series with n terms, $a = R$, and $r = 1 + i$, which after substituting into the formula for a geometric series and rearranging is

$$A_f = R\left(\frac{(1 + i)^n - 1}{i}\right).$$

Exercises 4–5

4. Write the sum without using summation notation, and find the sum.

    a. $\displaystyle\sum_{k=0}^{5} k$

    b. $\displaystyle\sum_{j=5}^{7} j^2$

    c. $\displaystyle\sum_{i=2}^{4} \frac{1}{i}$

5. Write each sum using summation notation. Do not evaluate the sum.

    a. $1^4 + 2^4 + 3^4 + 4^4 + 5^4 + 6^4 + 7^4 + 8^4 + 9^4$

    b. $1 + \cos(\pi) + \cos(2\pi) + \cos(3\pi) + \cos(4\pi) + \cos(5\pi)$

    c. $2 + 4 + 6 + \cdots + 1000$

Lesson Summary

- **SERIES**: Let $a_1, a_2, a_3, a_4, \ldots$ be a sequence of numbers. A sum of the form

$$a_1 + a_2 + a_3 + \cdots + a_n$$

for some positive integer n is called a *series* (or *finite series*) and is denoted S_n. The a_i's are called the *terms* of the series. The number S_n that the series adds to is called the *sum* of the series.

- **GEOMETRIC SERIES**: A *geometric series* is a series whose terms form a geometric sequence.

- **SUM OF A FINITE GEOMETRIC SERIES**: The sum S_n of the first n terms of the geometric series $S_n = a + ar + \cdots + ar^{n-1}$ (when $r \neq 1$) is given by

$$S_n = a\left(\frac{1 - r^n}{1 - r}\right).$$

  The *sum of a finite geometric series* can be written in summation notation as

$$\sum_{k=0}^{n-1} ar^k = a\left(\frac{1 - r^n}{1 - r}\right).$$

- The generic formula for calculating the future value of an annuity A_f in terms of the recurring payment R, interest rate i, and number of periods n is given by

$$A_f = R\left(\frac{(1 + i)^n - 1}{i}\right).$$

Problem Set

1. A car loan is one of the first secured loans most Americans obtain. Research used car prices and specifications in your area to find a reasonable used car that you would like to own (under $10,000). If possible, print out a picture of the car you selected.

   a. What is the year, make, and model of your vehicle?

   b. What is the selling price for your vehicle?

c.   The following table gives the monthly cost per $1,000 financed on a 5-year auto loan. Assume you have qualified for a loan with a 5% annual interest rate. What is the monthly cost of financing the vehicle you selected? (A formula is developed to find the monthly payment of a loan in Lesson 30.)

Five-Year (60-month) Loan	
Interest Rate	Amount per $1,000 Financed
1.0%	$17.09
1.5%	$17.31
2.0%	$17.53
2.5%	$17.75
3.0%	$17.97
3.5%	$18.19
4.0%	$18.41
4.5%	$18.64
5.0%	$18.87
5.5%	$19.10
6.0%	$19.33
6.5%	$19.56
7.0%	$19.80
7.5%	$20.04
8.0%	$20.28
8.5%	$20.52
9.0%	$20.76

d.   What is the gas mileage for your vehicle?

e.   Suppose that you drive 120 miles per week and gas costs $4 per gallon. How much does gas cost per month?

2.   Write the sum without using summation notation, and find the sum.

a. $\displaystyle\sum_{k=1}^{8} k$

b. $\displaystyle\sum_{k=-8}^{8} k$

c. $\displaystyle\sum_{k=1}^{4} k^3$

d. $\displaystyle\sum_{m=0}^{6} 2m$

e. $\displaystyle\sum_{m=0}^{6} 2m+1$

f. $\displaystyle\sum_{k=2}^{5} \frac{1}{k}$

g. $\displaystyle\sum_{j=0}^{3} (-4)^{j-2}$

h. $\displaystyle\sum_{m=1}^{4} 16\left(\frac{3}{2}\right)^m$

i. $\displaystyle\sum_{j=0}^{3} \frac{105}{2j+1}$

j. $\displaystyle\sum_{p=1}^{3} p \cdot 3^p$

k. $\displaystyle\sum_{j=1}^{6} 100$

l. $\displaystyle\sum_{k=0}^{4} \sin\left(\frac{k\pi}{2}\right)$

m. $\displaystyle\sum_{k=1}^{9} \log\left(\frac{k}{k+1}\right)$

(Hint: You do not need a calculator to find the sum.)

EUREKA
MATH™

3. Write the sum without using sigma notation. (You do not need to find the sum.)

   a. $\displaystyle\sum_{k=0}^{4} \sqrt{k+3}$

   b. $\displaystyle\sum_{i=0}^{8} x^i$

   c. $\displaystyle\sum_{j=1}^{6} jx^{j-1}$

   d. $\displaystyle\sum_{k=0}^{9} (-1)^k x^k$

4. Write each sum using summation notation.

   a. $1 + 2 + 3 + 4 + \cdots + 1000$

   b. $2 + 4 + 6 + 8 + \cdots + 100$

   c. $1 + 3 + 5 + 7 + \cdots + 99$

   d. $\dfrac{1}{2} + \dfrac{2}{3} + \dfrac{3}{4} + \cdots + \dfrac{99}{100}$

   e. $1^2 + 2^2 + 3^2 + 4^2 + \cdots + 10000^2$

   f. $1 + x + x^2 + x^3 + \cdots + x^{200}$

   g. $\dfrac{1}{1\cdot 2} + \dfrac{1}{2\cdot 3} + \dfrac{1}{3\cdot 4} + \cdots + \dfrac{1}{49\cdot 50}$

   h. $1\ln(1) + 2\ln(2) + 3\ln(3) + \cdots + 10\ln(10)$

5. Use the geometric series formulas to find the sum of the geometric series.

   a. $1 + 3 + 9 + \cdots + 2187$

   b. $1 + \dfrac{1}{2} + \dfrac{1}{4} + \dfrac{1}{8} + \cdots + \dfrac{1}{512}$

   c. $1 - \dfrac{1}{2} + \dfrac{1}{4} - \dfrac{1}{8} + \cdots - \dfrac{1}{512}$

   d. $0.8 + 0.64 + 0.512 + \cdots + 0.32768$

   e. $1 + \sqrt{3} + 3 + 3\sqrt{3} + \cdots + 243$

   f. $\displaystyle\sum_{k=0}^{5} 2^k$

   g. $\displaystyle\sum_{m=1}^{4} 5\left(\dfrac{3}{2}\right)^m$

   h. $1 - x + x^2 - x^3 + \cdots + x^{30}$ in terms of x

   i. $\displaystyle\sum_{m=0}^{11} 4^{\frac{m}{3}}$

j. $\sum_{n=0}^{14} \left(\sqrt[5]{6}\right)^n$

k. $\sum_{k=0}^{6} 2\left(\sqrt{3}\right)^k$

6. Let a_i represent the sequence of even natural numbers $\{2,4,6,8,\dots\}$ with $a_1 = 2$. Evaluate the following expressions.

   a. $\sum_{i=1}^{5} a_i$

   b. $\sum_{i=1}^{4} a_{2i}$

   c. $\sum_{i=1}^{5} (a_i - 1)$

7. Let a_i represent the sequence of integers giving the yardage gained per rush in a high school football game $\{3, -2, 17, 4, -8, 19, 2, 3, 3, 4, 0, 1, -7\}$.

   a. Evaluate $\sum_{i=1}^{13} a_i$. What does this sum represent in the context of the situation?

   b. Evaluate $\dfrac{\sum_{i=1}^{13} a_i}{13}$. What does this expression represent in the context of the situation?

   c. In general, if a_n describes any sequence of numbers, what does $\dfrac{\sum_{i=1}^{n} a_i}{n}$ represent?

8. Let b_n represent the sequence given by the following recursive formula: $b_1 = 10$, $b_n = 5b_{n-1}$.

   a. Write the first 4 terms of this sequence.

   b. Expand the sum $\sum_{i=1}^{4} b_i$. Is it easier to add this series, or is it easier to use the formula for the sum of a finite geometric sequence? Explain your answer. Evaluate $\sum_{i=1}^{4} b_i$.

   c. Write an explicit form for b_n.

   d. Evaluate $\sum_{i=1}^{10} b_i$.

9. Consider the sequence given by $a_1 = 20$, $a_n = \frac{1}{2} \cdot a_{n-1}$.

   a. Evaluate $\sum_{i=1}^{10} a_i$, $\sum_{i=1}^{100} a_i$, and $\sum_{i=1}^{1000} a_i$.

   b. What value does it appear this series is approaching as n continues to increase? Why might it seem like the series is bounded?

10. The sum of a geometric series with four terms is 60, and the common ratio is $r = \frac{1}{2}$. Find the first term.

11. The sum of the first four terms of a geometric series is 203, and the common ratio is 0.4. Find the first term.

12. The third term in a geometric series is $\dfrac{27}{2}$, and the sixth term is $\dfrac{729}{16}$. Find the common ratio.

13. The second term in a geometric series is 10, and the seventh term is 10,240. Find the sum of the first six terms.

14. Find the interest earned and the future value of an annuity with monthly payments of $200 for two years into an account that pays 6% interest per year compounded monthly.

15. Find the interest earned and the future value of an annuity with annual payments of $1,200 for 15 years into an account that pays 4% interest per year.

16. Find the interest earned and the future value of an annuity with semiannual payments of $1,000 for 20 years into an account that pays 7% interest per year compounded semiannually.

17. Find the interest earned and the future value of an annuity with weekly payments of $100 for three years into an account that pays 5% interest per year compounded weekly.

18. Find the interest earned and the future value of an annuity with quarterly payments of $500 for 12 years into an account that pays 3% interest per year compounded quarterly.

19. How much money should be invested every month with 8% interest per year compounded monthly in order to save up $10,000 in 15 months?

20. How much money should be invested every year with 4% interest per year in order to save up $40,000 in 18 years?

21. Julian wants to save up to buy a car. He is told that a loan for a car costs $274 a month for five years, but Julian does not need a car presently. He decides to invest in a structured savings plan for the next three years. Every month Julian invests $274 at an annual interest rate of 2% compounded monthly.

    a. How much will Julian have at the end of three years?

    b. What are the benefits of investing in a structured savings plan instead of taking a loan out? What are the drawbacks?

22. An *arithmetic series* is a series whose terms form an arithmetic sequence. For example, $2 + 4 + 6 + \cdots + 100$ is an arithmetic series since $2, 4, 6, 8, \ldots, 100$ is an arithmetic sequence with constant difference 2.

The most famous arithmetic series is $1 + 2 + 3 + 4 + \cdots + n$ for some positive integer n. We studied this series in Algebra I and showed that its sum is $S_n = \frac{n(n+1)}{2}$. It can be shown that the general formula for the sum of an arithmetic series $a + (a + d) + (a + 2d) + \cdots + [a + (n - 1)d]$ is

$$S_n = \frac{n}{2}[2a + (n - 1)d],$$

where a is the first term and d is the constant difference.

a. Use the general formula to show that the sum of $1 + 2 + 3 + \cdots + n$ is $S_n = \frac{n(n+1)}{2}$.

b. Use the general formula to find the sum of $2 + 4 + 6 + 8 + 10 + \cdots + 100$.

23. The sum of the first five terms of an arithmetic series is 25, and the first term is 2. Find the constant difference.

24. The sum of the first nine terms of an arithmetic series is 135, and the first term is 17. Find the ninth term.

25. The sum of the first and 100th terms of an arithmetic series is 101. Find the sum of the first 100 terms.

This page intentionally left blank

Lesson 30: Buying a Car

Classwork

Opening Exercise

Write a sum to represent the future amount of a structured savings plan (i.e., annuity) if you deposit $250 into an account each month for 5 years that pays 3.6% interest per year, compounded monthly. Find the future amount of your plan at the end of 5 years.

Example

Jack wanted to buy a $9,000 2-door sports coupe but could not pay the full price of the car all at once. He asked the car dealer if she could give him a loan where he paid a monthly payment. She told him she could give him a loan for the price of the car at an annual interest rate of 3.6% compounded monthly for 60 months (5 years).

The problems below exhibit how Jack's car dealer used the information above to figure out how much his monthly payment of R dollars per month should be.

a. First, the car dealer imagined how much she would have in an account if she deposited $9,000 into the account and left it there for 60 months at an annual interest rate of 3.6% compounded monthly. Use the compound interest formula $F = P(1 + i)^n$ to calculate how much she would have in that account after 5 years. This is the amount she would have in the account after 5 years if Jack gave her $9,000 for the car, and she immediately deposited it.

---

b. Next, she figured out how much would be in an account after 5 years if she took each of Jack's payments of R dollars and deposited it into a bank that earned 3.6% per year (compounded monthly). Write a sum to represent the future amount of money that would be in the annuity after 5 years in terms of R, and use the sum of a geometric series formula to rewrite that sum as an algebraic expression.

c. The car dealer then reasoned that, to be fair to her and Jack, the two final amounts in both accounts should be the same—that is, she should have the same amount in each account at the end of 60 months either way. Write an equation in the variable R that represents this equality.

d. She then solved her equation to get the amount R that Jack would have to pay monthly. Solve the equation in part (c) to find out how much Jack needed to pay each month.

Exercise

A college student wants to buy a car and can afford to pay $200 per month. If she plans to take out a loan at 6% interest per year with a recurring payment of $200 per month for four years, what price car can she buy?

Mathematical Modeling Exercise

In the Problem Set of Lesson 29, you researched the price of a car that you might like to own. In this exercise, we determine how much a car payment would be for that price for different loan options.

If you did not find a suitable car, select a car and selling price from the list below:

Car	Selling Price
2005 Pickup Truck	$9,000
2007 Two-Door Small Coupe	$7,500
2003 Two-Door Luxury Coupe	$1, 000
2006 Small SUV	$8,000
2008 Four-Door Sedan	$8,500

a. When you buy a car, you must pay sales tax and licensing and other fees. Assume that sales tax is 6% of the selling price and estimated license/title/fees are 2% of the selling price. If you put a $1,000 down payment on your car, how much money do you need to borrow to pay for the car and taxes and other fees?

b. Using the loan amount you computed above, calculate the monthly payment for the different loan options shown below:

Loan 1	36-month loan at 2%
Loan 2	48-month loan at 3%
Loan 3	60-month loan at 5%

c. Which plan, if any, keeps the monthly payment under $175? Of the plans under $175 per month, why might you choose a plan with fewer months even though it costs more per month?

Lesson Summary

The total cost of car ownership includes many different costs in addition to the selling price, such as sales tax, insurance, fees, maintenance, interest on loans, gasoline, etc.

The present value of an annuity formula can be used to calculate monthly loan payments given a total amount borrowed, the length of the loan, and the interest rate. The present value A_p (i.e., loan amount) of an annuity consisting of n recurring equal payments of size R and interest rate i per time period is

$$A_p = R\left(\frac{1-(1+i)^{-n}}{i}\right).$$

Amortization tables and online loan calculators can also help you plan for buying a car.

The amount of your monthly payment depends on the interest rate, the down payment, and the length of the loan.

Problem Set

1. Benji is 24 years old and plans to drive his new car about 200 miles per week. He has qualified for first-time buyer financing, which is a 60-month loan with 0% down at an interest rate of 4%. Use the information below to estimate the monthly cost of each vehicle.

    CAR A: 2010 Pickup Truck for $12,000, 22 miles per gallon

    CAR B: 2006 Luxury Coupe for $11,000, 25 miles per gallon

    Gasoline: $4.00 per gallon   New vehicle fees: $80   Sales Tax: 4.25%

    Maintenance Costs:

    (2010 model year or newer): 10% of purchase price annually

    (2009 model year or older): 20% of purchase price annually

    Insurance:

Average Rate Ages 25–29	$100 per month
If you are male	Add $10 per month
If you are female	Subtract $10 per month
Type of Car	
Pickup Truck	Subtract $10 per month
Small Two-Door Coupe or Four-Door Sedan	Subtract $10 per month
Luxury Two- or Four-Door Coupe	Add $15 per month
Ages 18–25	Double the monthly cost

    a. How much money will Benji have to borrow to purchase each car?

    b. What is the monthly payment for each car?

    c. What are the annual maintenance costs and insurance costs for each car?

    d. Which car should Benji purchase? Explain your choice.

2. Use the total initial cost of buying your car from the lesson to calculate the monthly payment for the following loan options.

Option	Number of Months	Down Payment	Interest Rate	Monthly Payment
Option A	48 months	$0	2.5%	
Option B	60 months	$500	3.0%	
Option C	60 months	$0	4.0%	
Option D	36 months	$1,000	0.9%	

   a. For each option, what is the total amount of money you will pay for your vehicle over the life of the loan?

   b. Which option would you choose? Justify your reasoning.

3. Many lending institutions allow you to pay additional money toward the principal of your loan every month. The table below shows the monthly payment for an $8,000 loan using Option A above if you pay an additional $25 per month.

Month/ Year	Payment	Principal Paid	Interest Paid	Total Interest	Balance
Aug. 2014	$200.31	$183.65	$16.67	$16.67	$7,816.35
Sept. 2014	$200.31	$184.03	$16.28	$32.95	$7,632.33
Oct. 2014	$200.31	$184.41	$15.90	$48.85	$7,447.91
Nov. 2014	$200.31	$184.80	$15.52	$64.37	$7,263.12
Dec. 2014	$200.31	$185.18	$15.13	$79.50	$7,077.94
Jan. 2015	$200.31	$185.57	$14.75	$94.25	$6,892.37
Feb. 2015	$200.31	$185.95	$14.36	$108.60	$6,706.42
Mar. 2015	$200.31	$186.34	$13.97	$122.58	$6,520.08
April 2015	$200.31	$186.73	$13.58	$136.16	$6,333.35
May 2015	$200.31	$187.12	$13.19	$149.35	$6,146.23
June 2015	$200.31	$187.51	$12.80	$162.16	$5,958.72
July 2015	$200.31	$187.90	$12.41	$174.57	$5,770.83
Aug. 2015	$200.31	$188.29	$12.02	$186.60	$5,582.54
Sept. 2015	$200.31	$188.68	$11.63	$198.23	$5,393.85
Oct. 2015	$200.31	$189.08	$11.24	$209.46	$5,204.78
Nov. 2015	$200.31	$189.47	$10.84	$220.31	$5,015.31
Dec. 2015	$200.31	$189.86	$10.45	$230.75	$4,825.45

Note: The months from January 2016 to December 2016 are not shown.

Jan. 2017	$ 200.31	$ 195.07	$ 5.24	$ 330.29	$ 2,320.92
Feb. 2017	$ 200.31	$ 195.48	$ 4.84	$ 335.12	$ 2,125.44
Mar. 2017	$ 200.31	$ 195.88	$ 4.43	$ 339.55	$ 1,929.56
April 2017	$ 200.31	$ 196.29	$ 4.02	$ 343.57	$ 1,733.27
May 2017	$ 200.31	$ 196.70	$ 3.61	$ 347.18	$ 1,536.57
June 2017	$ 200.31	$ 197.11	$ 3.20	$ 350.38	$ 1,339.45
July 2017	$ 200.31	$ 197.52	$ 2.79	$ 353.17	$ 1,141.93
Aug. 2017	$ 200.31	$ 197.93	$ 2.38	$ 355.55	$ 944.00
Sept. 2017	$ 200.31	$ 198.35	$ 1.97	$ 357.52	$ 745.65
Oct. 2017	$ 200.31	$ 198.76	$ 1.55	$ 359.07	$ 546.90
Nov. 2017	$ 200.31	$ 199.17	$ 1.14	$ 360.21	$ 347.72
Dec. 2017	$ 200.31	$ 199.59	$ 0.72	$ 360.94	$ 148.13
Jan. 2018	$ 148.44	$ 148.13	$ 0.31	$ 361.25	$ 0.00

How much money would you save over the life of an $8,000 loan using Option A if you paid an extra $25 per month compared to the same loan without the extra payment toward the principal?

4. Suppose you can afford only $200 a month in car payments and your best loan option is a 60-month loan at 3%. How much money could you spend on a car? That is, calculate the present value of the loan with these conditions.

5. Would it make sense for you to pay an additional amount per month toward your car loan? Use an online loan calculator to support your reasoning.

6. What is the sum of each series?

   a. $900 + 900(1.01)^1 + 900(1.01)^2 + \cdots + 900(1.01)^{59}$

   b. $\sum_{n=0}^{47} 15000 \left(1 + \frac{0.04}{12}\right)^n$

7. Gerald wants to borrow $12,000 in order to buy an engagement ring. He wants to repay the loan by making monthly installments for two years. If the interest rate on this loan is $9\frac{1}{2}\%$ per year, compounded monthly, what is the amount of each payment?

8. Ivan plans to surprise his family with a new pool using his Christmas bonus of $4,200 as a down payment. If the price of the pool is $9,500 and Ivan can finance it at an interest rate of $2\frac{7}{8}\%$ per year, compounded quarterly, how long is the term of the loan if his payments are $285.45 per quarter?

9. Jenny wants to buy a car by making payments of $120 per month for three years. The dealer tells her that she needs to put a down payment of $3,000 on the car in order to get a loan with those terms at a 9% interest rate per year, compounded monthly. How much is the car that Jenny wants to buy?

10. Kelsey wants to refinish the floors in her house and estimates that it will cost $39,000 to do so. She plans to finance the entire amount at $3\frac{1}{4}$% interest per year, compounded monthly for 10 years. How much is her monthly payment?

11. Lawrence coaches little league baseball and needs to purchase all new equipment for his team. He has $489 in donations, and the team's sponsor will take out a loan at $4\frac{1}{2}$% interest per year, compounded monthly for one year, paying up to $95 per month. What is the most that Lawrence can purchase using the donations and loan?

This page intentionally left blank

Lesson 31: Credit Cards

Classwork

Mathematical Modeling Exercise

You have charged \$1,500 for the down payment on your car to a credit card that charges 19.99% annual interest, and you plan to pay a fixed amount toward this debt each month until it is paid off. We denote the balance owed after the nth payment has been made as b_n.

a. What is the monthly interest rate, i? Approximate i to 5 decimal places.

b. You have been assigned to either the 50-team, the 100-team, or the 150-team, where the number indicates the size of the monthly payment R you make toward your debt. What is your value of R?

c. Remember that you can make any size payment toward a credit card debt, as long as it is at least as large as the minimum payment specified by the lender. Your lender calculates the minimum payment as the sum of 1% of the outstanding balance and the total interest that has accrued over the month, or \$25, whichever is greater. Under these stipulations, what is the minimum payment? Is your monthly payment R at least as large as the minimum payment?

d. Complete the following table to show 6 months of payments.

Month, n	Interest Due (in dollars)	Payment, R (in dollars)	Paid to Principal (in dollars)	Balance, b_n (in dollars)
0				1,500.00
1				
2				
3				
4				
5				
6				

e. Write a recursive formula for the balance b_n in month n in terms of the balance b_{n-1}.

f. Write an explicit formula for the balance b_n in month n, leaving the expression $1 + i$ in symbolic form.

g. Rewrite your formula in part (f) using r to represent the quantity $(1 + i)$.

h. What can you say about your formula in part (g)? What term do we use to describe r in this formula?

i. Write your formula from part (g) in summation notation using Σ.

EUREKA MATH

j. Apply the appropriate formula from Lesson 29 to rewrite your formula from part (g).

k. Find the month when your balance is paid off.

l. Calculate the total amount paid over the life of the debt. How much was paid solely to interest?

Problem Set

1. Suppose that you have a $2,000 balance on a credit card with a 29.99% annual interest rate, compounded monthly, and you can afford to pay $150 per month toward this debt.

   a. Find the amount of time it takes to pay off this debt. Give your answer in months and years.

   b. Calculate the total amount paid over the life of the debt.

   c. How much money was paid entirely to the interest on this debt?

2. Suppose that you have a $2,000 balance on a credit card with a 14.99% annual interest rate, and you can afford to pay $150 per month toward this debt.

   a. Find the amount of time it takes to pay off this debt. Give your answer in months and years.

   b. Calculate the total amount paid over the life of the debt.

   c. How much money was paid entirely to the interest on this debt?

3. Suppose that you have a $2,000 balance on a credit card with a 7.99% annual interest rate, and you can afford to pay $150 per month toward this debt.

   a. Find the amount of time it takes to pay off this debt. Give your answer in months and years.

   b. Calculate the total amount paid over the life of the debt.

   c. How much money was paid entirely to the interest on this debt?

4. Summarize the results of Problems 1, 2, and 3.

5. Brendan owes $1,500 on a credit card with an interest rate of 12%. He is making payments of $100 every month to pay this debt off. Maggie is also making regular payments to a debt owed on a credit card, and she created the following graph of her projected balance over the next 12 months.

   a. Who has the higher initial balance? Explain how you know.

   b. Who will pay their debt off first? Explain how you know.

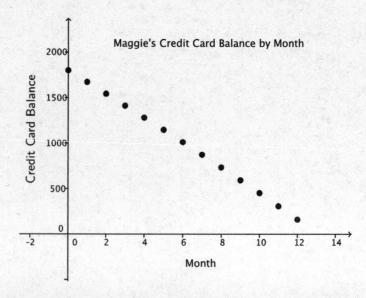

EUREKA
MATH™

6. Alan and Emma are both making $200 monthly payments toward balances on credit cards. Alan has prepared a table to represent his projected balances, and Emma has prepared a graph.

Alan's Credit Card Balance			
Month, n	Interest	Payment	Balance, b_n
0	——	——	2,000.00
1	41.65	200	1,841.65
2	38.35	200	1,680.00
3	34.99	200	1,514.99
4	31.55	200	1,346.54
5	28.04	200	1,174.58
6	24.46	200	999.04
7	20.81	200	819.85
8	17.07	200	636.92
9	13.26	200	450.18
10	9.37	200	259.55
11	5.41	200	64.96

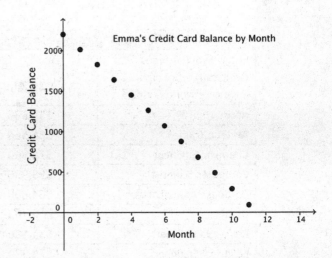

a. What is the annual interest rate on Alan's debt? Explain how you know.

b. Who has the higher initial balance? Explain how you know.

c. Who will pay their debt off first? Explain how you know.

d. What do your answers to parts (a), (b), and (c) tell you about the interest rate for Emma's debt?

7. Both Gary and Helena are paying regular monthly payments to a credit card balance. The balance on Gary's credit card debt can be modeled by the recursive formula $g_n = g_{n-1}(1.01666) - 200$ with $g_0 = 2500$, and the balance on Helena's credit card debt can be modeled by the explicit formula

$$h_n = 2000(1.01\,666)^n - 250\left(\frac{1.01666^n - 1}{0.01666}\right) \text{ for } n \geq 0.$$

a. Who has the higher initial balance? Explain how you know.

b. Who has the higher monthly payment? Explain how you know.

c. Who will pay their debt off first? Explain how you know.

8. In the next lesson, we will apply the mathematics we have learned to the purchase of a house. In preparation for that task, you need to come to class prepared with an idea of the type of house you would like to buy.

a. Research the median housing price in the county where you live or where you wish to relocate.

b. Find the range of prices that are within 25% of the median price from part (a). That is, if the price from part (a) was P, then your range is $0.75P$ to $1.25P$.

c. Look at online real estate websites, and find a house located in your selected county that falls into the price range specified in part (b). You will be modeling the purchase of this house in Lesson 32, so bring a printout of the real estate listing to class with you.

9. Select a career that interests you from the following list of careers. If the career you are interested in is not on this list, check with your teacher to obtain permission to perform some independent research. Once it has been selected, you will use the career to answer questions in Lesson 32 and Lesson 33.

Occupation	Median Starting Salary	Education Required
Entry-level full-time (waitstaff, office clerk, lawn care worker, etc.)	$20,200	High school diploma or GED
Accountant	$54,630	4-year college degree
Athletic Trainer	$36,560	4-year college degree
Chemical Engineer	$78,860	4-year college degree
Computer Scientist	$93,950	4-year college degree or more
Database Administrator	$64,600	4-year college degree
Dentist	$136,960	Graduate degree
Desktop Publisher	$34,130	4-year college degree
Electrical Engineer	$75,930	4-year college degree
Graphic Designer	$39,900	2- or 4-year college degree
HR Employment Specialist	$42,420	4-year college degree
HR Compensation Manager	$66,530	4-year college degree
Industrial Designer	$54,560	4-year college degree or more
Industrial Engineer	$68,620	4-year college degree
Landscape Architect	$55,140	4-year college degree
Lawyer	$102,470	Law degree
Occupational Therapist	$60,470	Master's degree
Optometrist	$91,040	Master's degree
Physical Therapist	$66,200	Master's degree
Physician—Anesthesiology	$259,948	Medical degree
Physician—Family Practice	$137,119	Medical degree
Physician's Assistant	$74,980	2 years college plus 2-year program
Radiology Technician	$47,170	2-year degree
Registered Nurse	$57,280	2- or 4-year college degree plus
Social Worker—Hospital	$48,420	Master's degree
Teacher—Special Education	$47,650	Master's degree
Veterinarian	$71,990	Veterinary degree

Lesson 32: Buying a House

Classwork

Mathematical Modeling Exercise

Now that you have studied the mathematics of structured savings plans, buying a car, and paying down a credit card debt, it's time to think about the mathematics behind the purchase of a house. In the Problem Set in Lesson 31, you selected a future career and a home to purchase. The question of the day is this: Can you buy the house you have chosen on the salary of the career you have chosen? You need to adhere to the following constraints:

- Mortgages are loans that are usually offered with 30-, 20-, or 15-year repayment options. Start with a 30-year mortgage.
- The annual interest rate for the mortgage will be 5%.
- Your payment includes the payment of the loan for the house and payments into an account called an *escrow account*, which is used to pay for taxes and insurance on your home. We approximate the annual payment to escrow as 1.2% of the home's selling price.
- The bank can only approve a mortgage if the total monthly payment for the house, including the payment to the escrow account, does not exceed 30% of your monthly salary.
- You have saved up enough money to put a 10% down payment on this house.

1. Will the bank approve a 30-year mortgage on the house that you have chosen?

2. Answer either (a) or (b) as appropriate.

    a. If the bank approves a 30-year mortgage, do you meet the criteria for a 20-year mortgage? If you could get a mortgage for any number of years that you want, what is the shortest term for which you would qualify?

    b. If the bank does not approve a 30-year mortgage, what is the maximum price of a house that fits your budget?

Problem Set

1. Use the house you selected to purchase in the Problem Set from Lesson 31 for this problem.

   a. What was the selling price of this house?

   b. Calculate the total monthly payment, R, for a 15-year mortgage at 5% annual interest, paying 10% as a down payment and an annual escrow payment that is 1.2% of the full price of the house.

2. In the summer of 2014, the average listing price for homes for sale in the Hollywood Hills was $2,663,995.

   a. Suppose you want to buy a home at that price with a 30-year mortgage at 5.25% annual interest, paying 10% as a down payment and with an annual escrow payment that is 1.2% of the full price of the home. What is your total monthly payment on this house?

   b. How much is paid in interest over the life of the loan?

3. Suppose that you would like to buy a home priced at $200,000. You plan to make a payment of 10% of the purchase price and pay 1.2% of the purchase price into an escrow account annually.

   a. Compute the total monthly payment and the total interest paid over the life of the loan for a 30-year mortgage at 4.8% annual interest.

   b. Compute the total monthly payment and the total interest paid over the life of the loan for a 20-year mortgage at 4.8% annual interest.

   c. Compute the total monthly payment and the total interest paid over the life of the loan for a 15-year mortgage at 4.8% annual interest.

4. Suppose that you would like to buy a home priced at $180,000. You qualify for a 30-year mortgage at 4.5% annual interest and pay 1.2% of the purchase price into an escrow account annually.

   a. Calculate the total monthly payment and the total interest paid over the life of the loan if you make a 3% down payment.

   b. Calculate the total monthly payment and the total interest paid over the life of the loan if you make a 10% down payment.

   c. Calculate the total monthly payment and the total interest paid over the life of the loan if you make a 20% down payment.

   d. Summarize the results of parts (a), (b), and (c) in the chart below.

Percent Down Payment	Amount of Down Payment	Total Interest Paid
3%		
10%		
20%		

5. The following amortization table shows the amount of payments to principal and interest on a $100,000 mortgage at the beginning and the end of a 30-year loan. These payments do not include payments to the escrow account.

Month/ Year	Payment	Principal Paid	Interest Paid	Total Interest	Balance
Sept. 2014	$ 477.42	$ 144.08	$ 333.33	$ 333.33	$ 99,855.92
Oct. 2014	$ 477.42	$ 144.56	$ 332.85	$ 666.19	$ 99,711.36
Nov. 2014	$ 477.42	$ 145.04	$ 332.37	$ 998.56	$ 99,566.31
Dec. 2014	$ 477.42	$ 145.53	$ 331.89	$ 1,330.45	$ 99,420.78
Jan. 2015	$ 477.42	$ 146.01	$ 331.40	$ 1,661.85	$ 99,274.77

⋮

Mar. 2044	$ 477.42	$ 467.98	$ 9.44	$ 71,845.82	$ 2,363.39
April 2044	$ 477.42	$ 469.54	$ 7.88	$ 71,853.70	$ 1,893.85
May 2044	$ 477.42	$ 471.10	$ 6.31	$ 71,860.01	$ 1,422.75
June 2044	$ 477.42	$ 472.67	$ 4.74	$ 71,864.75	$ 950.08
July 2044	$ 477.42	$ 474.25	$ 3.17	$ 71,867.92	$ 475.83
Aug. 2044	$ 477.42	$ 475.83	$ 1.59	$ 71,869.51	$ 0.00

a. What is the annual interest rate for this loan? Explain how you know.

b. Describe the changes in the amount of principal paid each month as the month n gets closer to 360.

c. Describe the changes in the amount of interest paid each month as the month n gets closer to 360.

6. Suppose you want to buy a $200,000 home with a 30-year mortgage at 4.5% annual interest paying 10% down with an annual escrow payment that is 1.2% of the price of the home.

    a. Disregarding the payment to escrow, how much do you pay toward the loan on the house each month?

    b. What is the total monthly payment on this house?

    c. The graph below depicts the amount of your payment from part (b) that goes to the interest on the loan and the amount that goes to the principal on the loan. Explain how you can tell which graph is which.

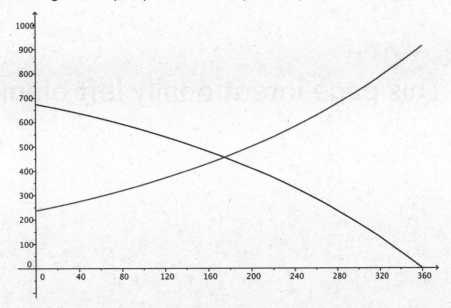

7. Student loans are very similar to both car loans and mortgages. The same techniques used for car loans and mortgages can be used for student loans. The difference between student loans and other types of loans is that usually students are not required to pay anything until 6 months after they stop being full-time students.

    a. An unsubsidized student loan will accumulate interest while a student remains in school. Sal borrows $9,000 his first term in school at an interest rate of 5.95% per year compounded monthly and never makes a payment. How much will he owe $4\frac{1}{2}$ years later? How much of that amount is due to compounded interest?

    b. If Sal pays the interest on his student loan every month while he is in school, how much money has he paid?

    c. Explain why the answer to part (a) is different than the answer to part (b).

8. Consider the sequence $a_0 = 10000$, $a_n = a_{n-1} \cdot \frac{1}{10}$ for $n \geq 1$.

    a. Write the explicit form for the nth term of the sequence.

    b. Evaluate $\sum_{k=0}^{4} a_k$.

    c. Evaluate $\sum_{k=0}^{6} a_k$.

    d. Evaluate $\sum_{k=0}^{8} a_k$ using the sum of a geometric series formula.

    e. Evaluate $\sum_{k=0}^{10} a_k$ using the sum of a geometric series formula.

    f. Describe the value of $\sum_{k=0}^{n} a_k$ for any value of $n \geq 4$.

This page intentionally left blank

Lesson 33: The Million Dollar Problem

Classwork

Opening Exercise

In Problem 1 of the Problem Set of Lesson 32, you calculated the monthly payment for a 15-year mortgage at a 5% annual interest rate for the house you chose. You need that monthly payment to answer these questions.

a. About how much do you expect your home to be worth in 15 years?

b. For $0 \leq x \leq 15$, plot the graph of the function $f(x) = P(1 + r)^x$ where r is the appreciation rate and P is the initial value of your home.

c. Compare the image of the graph you plotted in part (b) with a partner, and write your observations of the differences and similarities. What do you think is causing the differences that you see in the graphs? Share your observations with another group to see if your conclusions are correct.

Your friend Julia bought a home at the same time as you but chose to finance the loan over 30 years. Julia also was able to avoid a down payment and financed the entire value of her home. This allowed her to purchase a more expensive home, but 15 years later she still has not paid off the loan. Consider the following amortization table representing Julia's mortgage, and answer the following questions by comparing the table with your graph.

Payment #	Beginning Balance	Payment on Interest	Payment on Principal
1	$145,000	$543.75	$190.94
⋮	⋮	⋮	⋮
178	$96,784.14	$362.94	$371.75
179	$96,412.38	$361.55	$373.15
180	$96,039.23	$360.15	$374.55

d. In Julia's neighborhood, her home has grown in value at around 2.95% per year. Considering how much she still owes the bank, how much of her home does she own after 15 years (the equity in her home)? Express your answer in dollars to the nearest thousand and as a percent of the value of her home.

e. Reasoning from your graph in part (b) and the table above, if both you and Julia sell your homes in 15 years at the homes' appreciated values, who would have more equity?

f. How much more do you need to save over 15 years to have assets over $1,000,000?

Lesson 33:   The Million Dollar Problem

©2015 Great Minds. eureka-math.org
ALG II-M3-SE-B2-1.3.0-08.2015

Mathematical Modeling Exercises

Assume you can earn 7% interest annually, compounded monthly, in an investment account. Develop a savings plan so that you will have $1 million in assets in 15 years (including the equity in your paid-off house).

1. Use your answer to Opening Exercise, part (g) as the future value of your savings plan.

    a. How much will you have to save every month to save up $1 million in assets?

    b. Recall the monthly payment to pay off your home in 15 years (from Problem 1 of the Problem Set of Lesson 32). How much are the two together? What percentage of your monthly income is this for the profession you chose?

2. Write a report supported by the calculations you did above on how to save $1 million (or more) in your lifetime.

Problem Set

1. Consider the following scenario: You would like to save up $50,000 after 10 years and plan to set up a structured savings plan to make monthly payments at 4.125% interest annually, compounded monthly.

   a. What lump sum amount would you need to invest at this interest rate in order to have $50,000 after 10 years?

   b. Use an online amortization calculator to find the monthly payment necessary to take a loan for the amount in part (a) at this interest rate and for this time period.

   c. Use $A_f = R\left(\frac{(1+i)^n - 1}{i}\right)$ to solve for R.

   d. Compare your answers to part (b) and part (c). What do you notice? Why did this happen?

2. For structured savings plans, the future value of the savings plan as a function of the number of payments made at that point is an interesting function to examine. Consider a structured savings plan with a recurring payment of $450 made monthly and an annual interest rate of 5.875% compounded monthly.

   a. State the formula for the future value of this structured savings plan as a function of the number of payments made. Use f for the function name.

   b. Graph the function you wrote in part (a) for $0 \le x \le 216$.

   c. State any trends that you notice for this function.

   d. What is the approximate value of the function f for $x = 216$?

   e. What is the domain of f? Explain.

   f. If the domain of the function is restricted to natural numbers, is the function a geometric sequence? Why or why not?

   g. Recall that the nth partial sums of a geometric sequence can be represented with S_n. It is true that $f(x) = S_x$ for positive integers x, since it is a geometric sequence; that is, $S_x = \sum_{i=1}^{x} ar^i$. State the geometric sequence whose sums of the first x terms are represented by this function.

   h. April has been following this structured savings plan for 18 years. April says that taking out the money and starting over should not affect the total money earned because the interest rate does not change. Explain why April is incorrect in her reasoning.

3. Henry plans to have $195,000 in property in 14 years and would like to save up to $1 million by depositing $3,068.95 each month at 6% interest per year, compounded monthly. Tina's structured savings plan over the same time span is described in the following table:

   a. Who has the higher interest rate? Who pays more every month?

   b. At the end of 14 years, who has more money from their structured savings plan? Does this agree with what you expected? Why or why not?

   c. At the end of 40 years, who has more money from their structured savings plan?

Deposit #	Amount Saved
30	$110,574.77
31	$114,466.39
32	$118,371.79
33	$122,291.02
34	$126,224.14
⋮	⋮
167	$795,266.92
168	$801,583.49

4. Edgar and Paul are two brothers that both get an inheritance of $150,000. Both plan to save up over $1,000,000 in 25 years. Edgar takes his inheritance and deposits the money into an investment account earning 8% interest annually, compounded monthly, payable at the end of 25 years. Paul spends his inheritance but uses a structured savings plan that is represented by the sequence $b_n = 1\,275 + b_{n-1} \cdot \left(1 + \frac{0.0\,775}{12}\right)$ with $b_0 = 1\,275$ in order to save up over $1,000,000.

    a. Which of the two has more money at the end of 25 years?

    b. What are the pros and cons of both brothers' plans? Which would you rather do? Why?

This page intentionally left blank

Eureka Math
Algebra II
Module 4

Special thanks go to the Gordan A. Cain Center and to the Department of Mathematics at Louisiana State University for their support in the development of *Eureka Math*.

Lesson 1: Chance Experiments, Sample Spaces, and Events

Classwork

Alan is designing a probability game. He plans to present the game to people who will consider financing his idea. Here is a description of the game:

- The game includes the following materials:
  - A fair coin with a head and a tail
  - Spinner 1 with three equal area sectors identified as 1, 2, and 3
  - Spinner 2 with six equal area sectors identified as 1, 2, 3, 4, 5, and 6
  - A card bag containing six cards. Four cards are blue with the letter *A* written on one card, *B* on another card, *C* on a third card, and *D* on the fourth card. Two cards are red with the letter *E* written on one card and the letter *F* written on the other. (Although actually using colored paper is preferable, slips of paper with the words *blue* or *red* written will also work.)
  - A set of scenario cards, each describing a chance experiment and a set of five possible events based on the chance experiment

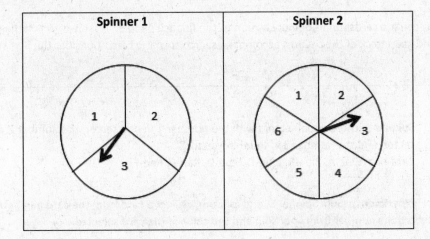

Card Bag:

Blue A	Blue B	Blue C	Blue D	Red E	Red F

- The game is played by two players (or two small groups of players) identified as Player 1 and Player 2.
- Rules of the game:
  - The scenario cards are shuffled, and one is selected.
  - Each player reads the description of the chance experiment and the description of the five possible outcomes.
  - Players independently assign the numbers 1–5 (no repeats) to the five events described on the scenario card based on how likely they think the event is to occur, with 5 being most likely and 1 being least likely.
  - Once players have made their assignments, the chance experiment described on the scenario card is performed. Points are then awarded based on the outcome of the chance experiment. If the event described on the scenario card has occurred, the player earns the number of points corresponding to the number that player assigned to that event (1–5 points). If an event occurs that is not described on the scenario card, then no points are awarded for that event.
  - If an outcome is described by two or more events on the scenario card, the player selects the higher point value.
  - The chance experiment is repeated four more times with points being awarded each time the chance experiment is performed.
  - The player with the largest number of points at the end of the game is the winner.

Alan developed two scenario cards for his demonstration to the finance people. A table in which the players can make their assignments and keep track of their scores accompanies each scenario card. Consider the first scenario card Alan developed.

Scenario Card 1

Game Tools: Spinner 1 (three equal sectors with the number 1 in one sector, the number 2 in the second sector, and the number 3 in the third sector)
Card bag (Blue-A, Blue-B, Blue-C, Blue-D, Red-E, Red-F)

Directions (chance experiment): Spin Spinner 1, and randomly select a card from the card bag (four blue cards and two red cards). Record the number from your spin and the color of the card selected.

Five Events of Interest:

Outcome is an odd number on Spinner 1 and a red card from the card bag.	Outcome is an odd number on Spinner 1.	Outcome is an odd number on Spinner 1 and a blue card from the card bag.	Outcome is an even number from Spinner 1 or a red card from the card bag.	Outcome is not a blue card from the card bag.

Lesson 1:   Chance Experiments, Sample Spaces, and Events

Player:

Scoring Card for Scenario 1:

Turn	Outcome from Spinner 1	Outcome from the Card Bag	Points
1			
2			
3			
4			
5			

Here is an example of Alan demonstrating the first scenario card. The chance experiment for Scenario Card 1 is "Spin Spinner 1, and record the number. Randomly select a card from the card bag (four blue cards and two red cards). Record the color of the card selected."

Alan assigned the numbers 1–5 to the descriptions, as shown below. Once a number is assigned, it cannot be used again.

Five Events of Interest:

Outcome is an odd number on Spinner 1 and a red card from the card bag.	Outcome is an odd number on Spinner 1.	Outcome is an odd number on Spinner 1 and a blue card from the card bag.	Outcome is an even number from Spinner 1 or a red card from the card bag.	Outcome is not a blue card from the card bag.
3	1	4	2	5

Alan is now ready to take his five turns. The results were recorded from the spinner and the card bag. Based on the results, Alan earned the points indicated for each turn.

Player: Player 1

Scoring Card for Scenario 1:

Turn	Outcome from Spinner 1	Outcome from the Card Bag	Points Based on Alan's Assignment of the Numbers to the Five Events
1	2	Blue	2
2	1	Red	5
3	1	Red	5
4	3	Blue	4
5	2	Blue	2

Alan earned a total of 18 points. The game now turns to Player 2. Player 2 assigns the numbers 1–5 to the same description of outcomes. Player 2 does not have to agree with the numbers Alan assigned. After five turns, the player with the most number of points is the winner.

Exploratory Challenge/Exercises 1–13

1. Would you change any of the assignments of 1–5 that Alan made? Explain your answer. Assign the numbers 1–5 to the event descriptions based on what you think is the best strategy to win the game.

Outcome is an odd number on Spinner 1 and a red card from the card bag.	Outcome is an odd number on Spinner 1.	Outcome is an odd number on Spinner 1 and a blue card from the card bag.	Outcome is an even number from Spinner 1 or a red card from the card bag.	Outcome is not a blue card from the card bag.

2. Carry out a turn by observing an outcome from spinning Spinner 1 and picking a card. How many points did you earn from this first turn?

3. Complete four more turns (for a total of five), and determine your final score.

   Player: Your Turn

   Scoring Card for Scenario 1:

Trial	Outcome from Spinner 1	Outcome from the Card Bag	Points Based on Your Assignment of Numbers to the Events
1			
2			
3			
4			
5			

4. If you changed the numbers assigned to the descriptions, was your score better than Alan's score? Did you expect your score to be better? Explain. If you did not change the numbers from those that Alan assigned, explain why you did not change them.

5. Spinning Spinner 1 and drawing a card from the card bag is a *chance experiment*. One possible outcome of this experiment is (1, Blue-A). Recall that the *sample space* for a chance experiment is the set of all possible outcomes. What is the sample space for the chance experiment of Scenario Card 1?

6. Are the outcomes in the sample space equally likely? Explain your answer.

7. Recall that an *event* is a collection of outcomes from the sample space. One event of interest for someone with Scenario Card 1 is "odd number on Spinner 1 and a red card." What are the outcomes that make up this event? List the outcomes of this event in the first row of Table 1 (see Exercise 9).

8. What is the probability of getting an odd number on Spinner 1 and picking a red card from the card bag? Also enter this probability in Table 1 (see Exercise 9).

9. Complete Table 1 by listing the outcomes for the other events and their probabilities based on the chance experiment for this scenario card.

Table 1

Event	Outcomes	Probability
Odd number on Spinner 1 and a red card from the card bag		
Odd number on Spinner 1		
Odd number on Spinner 1 and a blue card from the card bag		
Even number on Spinner 1 or a red card from the card bag		
Not picking a blue card from the card bag		

10. Based on the above probabilities, how would you assign the numbers 1–5 to each of the game descriptions? Explain.

EUREKA MATH™

11. If you changed any of the points assigned to the game descriptions, play the game again at least three times, and record your final scores for each game. Do you think you have the best possible assignment of numbers to the events for this scenario card? If you did not change the game descriptions, also play the game so that you have at least three final scores. Compare your scores with scores of other members of your class. Do you think you have the best assignment of numbers to the events for this scenario card?

Turn	Outcome from Spinner 1	Outcome from the Card Bag	Points Based on the Assignment of Points in Exercise 10
1			
2			
3			
4			
5			

12. Why might you not be able to answer the question of whether or not you have the best assignment of numbers to the game descriptions with at least three final scores?

13. Write your answers to the following questions independently, and then share your responses with a neighbor:

   a. How did you make decisions about what to bet on?

   b. How do the ideas of probability help you make decisions?

Lesson Summary

- **SAMPLE SPACE:** The *sample space* of a chance experiment is the collection of all possible outcomes for the experiment.
- **EVENT:** An *event* is a collection of outcomes of a chance experiment.
- For a chance experiment in which outcomes of the sample space are equally likely, the probability of an event is the number of outcomes in the event divided by the number of outcomes in the sample space.
- Some events are described in terms of *or*, *and*, or *not*.

Problem Set

Consider a second scenario card that Alan created for his game:

Scenario Card 2

Tools: **Spinner 1**
Spinner 2: a spinner with six equal sectors (Place the number 1 in a sector, the number 2 in a second sector, the number 3 in a third sector, the number 4 in a fourth sector, the number 5 in a fifth sector, and the number 6 in the last sector.)

Directions (chance experiment): Spin Spinner 1, and spin Spinner 2. Record the number from Spinner 1, and record the number from Spinner 2.

Five Events of Interest:

Outcome is an odd number on Spinner 2.	Outcome is an odd number on Spinner 1 and an even number on Spinner 2.	Outcome is the sum of 7 from the numbers received from Spinner 1 and Spinner 2.	Outcome is an even number on Spinner 2.	Outcome is the sum of 2 from the numbers received from Spinner 1 and Spinner 2.

Player:
Scoring Card for Scenario 2:

Turn	Outcome from Spinner 1	Outcome from Spinner 2	Points
1			
2			
3			
4			
5			

1. Prepare Spinner 1 and Spinner 2 for the chance experiment described on this second scenario card. (Recall that Spinner 2 has six equal sectors.)

2. What is the sample space for the chance experiment described on this scenario card?

3. Based on the sample space, determine the outcomes and the probabilities for each of the events on this scenario card. Complete the table below.

Event	Outcomes	Probability
Outcome is an odd number on Spinner 2.		
Outcome is an odd number on Spinner 1 and an even number on Spinner 2.		
Outcome is the sum of 7 from the numbers received from Spinner 1 and Spinner 2.		
Outcome is an even number on Spinner 2.		
Outcome is the sum of 2 from the numbers received from Spinner 1 and Spinner 2.		

4. Assign the numbers 1–5 to the events described on the scenario card.

   Five Events of Interest: Scenario 2

Outcome is an odd number on Spinner 2.	Outcome is an odd number on Spinner 1 and an even number on Spinner 2.	Outcome is the sum of 7 from the numbers received from Spinner 1 and Spinner 2.	Outcome is an even number on Spinner 2.	Outcome is the sum of 2 from the numbers received from Spinner 1 and Spinner 2.

EUREKA
MATH™

5. Determine at least three final scores based on the numbers you assigned to the events.

Player: Scott

Trial	Outcome from Spinner 1	Outcome from Spinner 2	Points (see Problem 4)
1			
2			
3			
4			
5			

Player: Scott

Trial	Outcome from Spinner 1	Outcome from Spinner 2	Points (see Problem 4)
1			
2			
3			
4			
5			

Player: Scott

Trial	Outcome from Spinner 1	Outcome from Spinner 2	Points (see Problem 4)
1			
2			
3			
4			
5			

EUREKA
MATH™

6. Alan also included a fair coin as one of the scenario tools. Develop a scenario card (Scenario Card 3) that uses the coin and one of the spinners. Include a description of the chance experiment and descriptions of five events relevant to the chance experiment.

Scenario Card 3

Tools: **Fair coin (head or tail)**
                    Spinner 1

Directions (chance experiment):

Five Events of Interest:

---

7. Determine the sample space for your chance experiment. Then, complete the table below for the five events on your scenario card. Assign the numbers 1–5 to the descriptions you created.

Event	Outcomes	Probability

8. Determine a final score for your game based on five turns.

Turn			Points
1			
2			
3			
4			
5			

Lesson 1:   Chance Experiments, Sample Spaces, and Events

EUREKA
MATH™

Lesson 2: Calculating Probabilities of Events Using Two-Way Tables

Classwork

Example 1: Building a New High School

The school board of Waldo, a rural town in the Midwest, is considering building a new high school primarily funded by local taxes. They decided to interview eligible voters to determine if the school board should build a new high school facility to replace the current high school building. There is only one high school in the town. Every registered voter in Waldo was interviewed. In addition to asking about support for a new high school, data on gender and age group were also recorded. The data from these interviews are summarized below.

Age (in years)	Should Our Town Build a New High School?					
	Yes		No		No Answer	
	Male	Female	Male	Female	Male	Female
18–25	29	32	8	6	0	0
26–40	53	60	40	44	2	4
41–65	30	36	44	35	2	2
66 and Older	7	26	24	29	2	0

Exercises 1–8: Building a New High School

1. Based on this survey, do you think the school board should recommend building a new high school? Explain your answer.

2. An eligible voter is picked at random. If this person is 21 years old, do you think he would indicate that the town should build a high school? Why or why not?

3. An eligible voter is picked at random. If this person is 55 years old, do you think she would indicate that the town should build a high school? Why or why not?

4. The school board wondered if the probability of recommending a new high school was different for different age categories. Why do you think the survey classified voters using the age categories 18–25 years old, 26–40 years old, 41–65 years old, and 66 years old and older?

5. It might be helpful to organize the data in a two-way frequency table. Use the given data to complete the following two-way frequency table. Note that the age categories are represented as rows, and the possible responses are represented as columns.

	Yes	No	No Answer	Total
18–25 Years Old				
26–40 Years Old				
41–65 Years Old				
66 Years Old and Older				
Total				

6. A local news service plans to write an article summarizing the survey results. Three possible headlines for this article are provided below. Is each headline accurate or inaccurate? Support your answer using probabilities calculated using the table above.

Headline 1: *Waldo Voters Likely to Support Building a New High School*

Headline 2: *Older Voters Less Likely to Support Building a New High School*

Headline 3: *Younger Voters Not Interested in Building a New High School*

EUREKA
MATH™

7. The school board decided to put the decision on whether or not to build the high school up for a referendum in the next election. At the last referendum regarding this issue, only 25 of the eligible voters ages 18–25 voted, 110 of the eligible voters ages 26–40 voted, 130 of the eligible voters ages 41–65 voted, and 80 of the eligible voters ages 66 and older voted. If the voters in the next election turn out in similar numbers, do you think this referendum will pass? Justify your answer.

8. Is it possible that your prediction of the election outcome might be incorrect? Explain.

Example 2: Smoking and Asthma

Health officials in Milwaukee, Wisconsin, were concerned about teenagers with asthma. People with asthma often have difficulty with normal breathing. In a local research study, researchers collected data on the incidence of asthma among students enrolled in a Milwaukee public high school.

Students in the high school completed a survey that was used to begin this research. Based on this survey, the probability of a randomly selected student at this high school having asthma was found to be 0.193. Students were also asked if they had at least one family member living in their house who smoked. The probability of a randomly selected student having at least one member in his (or her) household who smoked was reported to be 0.421.

Exercises 9–14

It would be easy to calculate probabilities if the data for the students had been organized into a two-way table like the one used in Exercise 5. But there is no table here, only probability information. One way around this is to think about what the table might have been if there had been 1,000 students at the school when the survey was given. This table is called a *hypothetical 1000 two-way table*.

What if the population of students at this high school was 1,000? The population was probably not exactly 1,000 students, but using an estimate of 1,000 students provides an easier way to understand the given probabilities. Connecting these estimates to the actual population is completed in a later exercise. Place the value of 1,000 in the cell representing the total population. Based on a hypothetical 1000 population, consider the following table:

	No Household Member Smokes	At Least One Household Member Smokes	Total
Student Has Asthma	Cell 1	Cell 2	Cell 3
Student Does Not Have Asthma	Cell 4	Cell 5	Cell 6
Total	Cell 7	Cell 8	1,000

9. The probability that a randomly selected student at this high school has asthma is 0.193. This probability can be used to calculate the value of one of the cells in the table above. Which cell is connected to this probability? Use this probability to calculate the value of that cell.

10. The probability that a randomly selected student has at least 1 household member who smokes is 0.421. Which cell is connected to this probability? Use this probability to calculate the value of that cell.

11. In addition to the previously given probabilities, the probability that a randomly selected student has at least one household member who smokes and has asthma is 0.120. Which cell is connected to this probability? Use this probability to calculate the value of that cell.

12. Complete the two-way frequency table by calculating the values of the other cells in the table.

	No Household Member Smokes	At Least One Household Member Smokes	Total
Student Has Asthma			
Student Does Not Have Asthma			
Total			1,000

EUREKA
MATH™

13. Based on your completed two-way table, estimate the following probabilities as a fraction and also as a decimal (rounded to three decimal places):

   a.   A randomly selected student has asthma. What is the probability this student has at least 1 household member who smokes?

   b.   A randomly selected student does not have asthma. What is the probability this student has at least one household member who smokes?

   c.   A randomly selected student has at least one household member who smokes. What is the probability this student has asthma?

14. Do you think that whether or not a student has asthma is related to whether or not this student has at least one family member who smokes? Explain your answer.

Lesson Summary

Data organized in a two-way frequency table can be used to calculate probabilities.

In certain problems, probabilities that are known can be used to create a hypothetical 1000 two-way table. The hypothetical population of 1,000 can then be used to calculate probabilities.

Probabilities are always interpreted in context.

Problem Set

1. The Waldo School Board asked eligible voters to evaluate the town's library service. Data are summarized in the following table:

Age (in years)	How Would You Rate Our Town's Library Services?							
	Good		Average		Poor		Do Not Use Library	
	Male	Female	Male	Female	Male	Female	Male	Female
18–25	10	8	5	7	5	5	17	18
26–40	30	28	25	30	20	30	20	20
41–65	30	32	26	21	15	10	5	10
66 and Older	21	25	8	15	2	10	2	5

a. What is the probability that a randomly selected person who completed the survey rated the library as good?

b. Imagine talking to a randomly selected male voter who had completed the survey. How do you think this person rated the library services? Explain your answer.

c. Use the given data to construct a two-way table that summarizes the responses on gender and rating of the library services. Use the following template as your guide:

	Good	Average	Poor	Do Not Use	Total
Male					
Female					
Total					

d. Based on your table, answer the following:
   i.   A randomly selected person who completed the survey is male. What is the probability he rates the library services as good?
   ii. A randomly selected person who completed the survey is female. What is the probability she rates the library services as good?

e. Based on your table, answer the following:
   i.   A randomly selected person who completed the survey rated the library services as good. What is the probability this person is male?
   ii. A randomly selected person who completed the survey rated the library services as good. What is the probability this person is female?

f. Do you think there is a difference in how male and female voters rated library services? Explain your answer.

2. Obedience School for Dogs is a small franchise that offers obedience classes for dogs. Some people think that larger dogs are easier to train and, therefore, should not be charged as much for the classes. To investigate this claim, dogs enrolled in the classes were classified as large (30 pounds or more) or small (under 30 pounds). The dogs were also classified by whether or not they passed the obedience class offered by the franchise. 45% of the dogs involved in the classes were large. 60% of the dogs passed the class. Records indicate that 40% of the dogs in the classes were small and passed the course.

   a. Complete the following hypothetical 1000 two-way table:

	Passed the Course	Did Not Pass the Course	Total
Large Dogs			
Small Dogs			
Total			

   b. Estimate the probability that a dog selected at random from those enrolled in the classes passed the course.

   c. A dog was randomly selected from the dogs that completed the class. If the selected dog was a large dog, what is the probability this dog passed the course?

   d. A dog was randomly selected from the dogs that completed the class. If the selected dog is a small dog, what is the probability this dog passed the course?

   e. Do you think dog size and whether or not a dog passes the course are related?

   f. Do you think large dogs should get a discount? Explain your answer.

This page intentionally left blank

Lesson 3: Calculating Conditional Probabilities and Evaluating Independence Using Two-Way Tables

Classwork

Example 1

Students at Rufus King High School were discussing some of the challenges of finding space for athletic teams to practice after school. Part of the problem, according to Kristin, is that female students are more likely to be involved in an after-school athletics program than male students. However, the athletic director assigns the available facilities as if male students are more likely to be involved. Before suggesting changes to the assignments, the students decided to investigate.

Suppose the following information is known about Rufus King High School: 40% of students are involved in one or more of the after-school athletics programs offered at the school. It is also known that 58% of the school's students are female. The students decide to construct a hypothetical 1000 two-way table, like Table 1, to organize the data.

Table 1: Participation in After-School Athletics Programs (Yes or No) by Gender

	Yes—Participate in After-School Athletics Programs	No—Do Not Participate in After-School Athletics Programs	Total
Female	Cell 1	Cell 2	Cell 3
Male	Cell 4	Cell 5	Cell 6
Total	Cell 7	Cell 8	Cell 9

Exercises 1–6: Organizing the Data

1. What cell in Table 1 represents a hypothetical group of 1,000 students at Rufus King High School?

2. What cells in Table 1 can be filled based on the information given about the student population? Place these values in the appropriate cells of the table based on this information.

3. Based only on the cells you completed in Exercise 2, which of the following probabilities can be calculated, and which cannot be calculated? Calculate the probability if it can be calculated. If it cannot be calculated, indicate why.

    a. The probability that a randomly selected student is female

    b. The probability that a randomly selected student participates in an after-school athletics program

    c. The probability that a randomly selected student who does not participate in an after-school athletics program is male

    d. The probability that a randomly selected male student participates in an after-school athletics program

4. The athletic director indicated that 23.2% of the students at Rufus King are female and participate in after-school athletics programs. Based on this information, complete Table 1.

5. Consider the cells 1, 2, 4, and 5 of Table 1. Identify which of these cells represent students who are female or who participate in after-school athletics programs.

6. What cells of the two-way table represent students who are male and do not participate in after-school athletics programs?

Lesson 3:   Calculating Conditional Probabilities and Evaluating Independence
          Using Two-Way Tables

EUREKA
MATH™

Example 2

The completed hypothetical 1000 table organizes information in a way that makes it possible to answer various questions. For example, you can investigate whether female students are more likely to be involved in the after-school athletic programs.

Consider the following events:

- Let A represent the event "a randomly selected student is female."
- Let "not A" represent "the *complement* of A." The complement of A represents the event "a randomly selected student is not female," which is equivalent to the event "a randomly selected student is male."
- Let B represent the event "a randomly selected student participates in an after-school athletics program."
- Let "not B" represent "the *complement* of B." The complement of B represents the event "a randomly selected student does not participate in an after-school athletics program."
- Let "A or B" (described as A *union* B) represent the event "a randomly selected student is female or participates in an after-school athletics program."
- Let "A and B" (described as A *intersect* B) represent the event "a randomly selected student is female and participates in an after-school athletics program."

Exercises 7–9

7. Based on the descriptions above, describe the following events in words:

   a. Not A or not B

   b. A and not B

8. Based on the above descriptions and Table 1, determine the probability of each of the following events:

   a. A

   b. B

   c. Not A

d.   Not B

e. A or B

f. A and B

9.   Determine the following values:

a.   The probability of A plus the probability of not A

b.   The probability of B plus the probability of not B

c.   What do you notice about the results of parts (a) and (b)? Explain.

Example 3: Conditional Probability

Another type of probability is called a *conditional probability*. Pulling apart the two-way table helps to define a conditional probability.

	Yes—Participate in After-School Athletics Program	No—Do Not Participate in After-School Athletics Program	Total
Female	Cell 1	Cell 2	Cell 3

Suppose that a randomly selected student is female. What is the probability that the selected student participates in an after-school athletics program? This probability is an example of what is called a *conditional probability*. This probability is calculated as the number of students who are female and participate in an after-school athletics program (or the students in cell 1) divided by the total number of female students (or the students in cell 3).

EUREKA
MATH™

Exercises 10–15

10. The following are also examples of conditional probabilities. Answer each question.

   a. What is the probability that if a randomly selected student is female, she participates in the after-school athletic program?

   b. What is the probability that if a randomly selected student is female, she does not participate in after-school athletics?

11. Describe two conditional probabilities that can be determined from the following row in Table 1:

	Yes—Participate in After-School Athletics Program	No—Do Not Participate in After-School Athletics Program	Total
Male	Cell 4	Cell 5	Cell 6

12. Describe two conditional probabilities that can be determined from the following column in Table 1:

	Yes—Participate in After-School Athletics Program
Female	Cell 1
Male	Cell 4
Total	Cell 7

13. Determine the following conditional probabilities:

    a.   A randomly selected student is female. What is the probability she participates in an after-school athletics program? Explain how you determined your answer.

    b.   A randomly selected student is male. What is the probability he participates in an after-school athletics program?

    c.   A student is selected at random. What is the probability this student participates in an after-school athletics program?

14. Based on the answers to Exercise 13, do you think that female students are more likely to be involved in after-school athletics programs? Explain your answer.

15. What might explain the concern female students expressed in the beginning of this lesson about the problem of assigning practice space?

EUREKA
MATH™

Lesson Summary

Data organized in a two-way frequency table can be used to calculate probabilities. Two-way frequency tables can also be used to calculate conditional probabilities.

In certain problems, probabilities that are known can be used to create a hypothetical 1000 two-way table. This hypothetical population of 1,000 can be used to calculate conditional probabilities.

Probabilities are always interpreted by the context of the data.

Problem Set

Oostburg College has a rather large marching band. Engineering majors were heard bragging that students majoring in engineering are more likely to be involved in the marching band than students from other majors.

1. If the above claim is accurate, does that mean that most of the band is engineering students? Explain your answer.

2. The following graph was prepared to investigate the above claim:

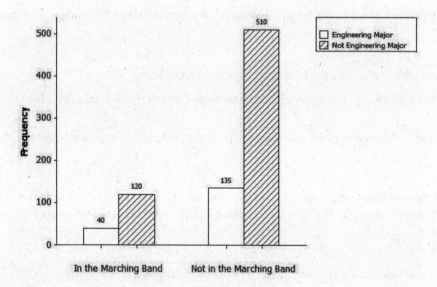

Based on the graph, complete the following two-way frequency table:

	In the Marching Band	Not in the Marching Band	Total
Engineering Major			
Not Engineering Major			
Total			

---

3. Let M represent the event that a randomly selected student is in the marching band. Let E represent the event that a randomly selected student is an engineering major.

   a. Describe the event represented by the complement of M.

   b. Describe the event represented by the complement of E.

   c. Describe the event M and E (M intersect E).

   d. Describe the event M or E (M union E).

4. Based on the completed two-way frequency table, determine the following, and explain how you got your answer:

   a. The probability that a randomly selected student is in the marching band

   b. The probability that a randomly selected student is an engineering major

   c. The probability that a randomly selected student is in the marching band and an engineering major

   d. The probability that a randomly selected student is in the marching band and not an engineering major

5. Indicate if the following conditional probabilities would be calculated using the rows or the columns of the two-way frequency table:

   a. A randomly selected student is majoring in engineering. What is the probability this student is in the marching band?

   b. A randomly selected student is not in the marching band. What is the probability that this student is majoring in engineering?

6. Based on the two-way frequency table, determine the following conditional probabilities:

   a. A randomly selected student is majoring in engineering. What is the probability that this student is in the marching band?

   b. A randomly selected student is not majoring in engineering. What is the probability that this student is in the marching band?

7. The claim that started this investigation was that students majoring in engineering are more likely to be in the marching band than students from other majors. Describe the conditional probabilities that would be used to determine if this claim is accurate.

8. Based on the two-way frequency table, calculate the conditional probabilities identified in Problem 7.

9. Do you think the claim that students majoring in engineering are more likely to be in the marching band than students for other majors is accurate? Explain your answer.

10. There are 40 students at Oostburg College majoring in computer science. Computer science is not considered an engineering major. Calculate an estimate of the number of computer science majors you think are in the marching band. Explain how you calculated your estimate.

Lesson 4: Calculating Conditional Probabilities and Evaluating Independence Using Two-Way Tables

Classwork

Exercise 1

In previous lessons, conditional probabilities were used to investigate whether or not there is a connection between two events. This lesson formalizes this idea and introduces the concept of *independence*.

1. Several questions are posed below. Each question is about a possible connection between two events. For each question, identify the two events, and indicate whether or not you think that there would be a connection. Explain your reasoning.

   a. Are high school students whose parents or guardians set a midnight curfew less likely to have a traffic violation than students whose parents or guardians have not set such a curfew?

   b. Are left-handed people more likely than right-handed people to be interested in the arts?

   c. Are students who regularly listen to classical music more likely to be interested in mathematics than students who do not regularly listen to classical music?

   d. Are people who play video games more than 10 hours per week more likely to select football as their favorite sport than people who do not play video games more than 10 hours per week?

Two events are independent when knowing that one event has occurred does not change the likelihood that the second event has occurred. How can conditional probabilities be used to tell if two events are independent or not independent?

Exercises 2–6

Recall the hypothetical 1000 two-way frequency table that was used to classify students at Rufus King High School according to gender and whether or not they participated in an after-school athletics program.

Table 1: Participation of Male and Female Students in an After-School Athletics Program

	Participate in an After-School Athletics Program	Do Not Participate in an After-School Athletics Program	Total
Female			
Male			
Total			

2. For each of the following, indicate whether the probability described is one that can be calculated using the values in Table 1. Also indicate whether or not it is a conditional probability.

   a. The probability that a randomly selected student participates in an after-school athletics program

   b. The probability that a randomly selected student who is female participates in an after-school athletics program

   c. The probability that a randomly selected student who is male participates in an after-school athletics program

3. Use Table 1 to calculate each of the probabilities described in Exercise 2.

   a. The probability that a randomly selected student participates in an after-school athletics program

Calculating Conditional Probabilities and Evaluating Independence Using Two-Way Tables

b. The probability that a randomly selected student who is female participates in an after-school athletics program

c. The probability that a randomly selected student who is male participates in an after-school athletics program

4. Would your prediction of whether or not a student participates in an after-school athletics program change if you knew the gender of the student? Explain your answer.

Two events are *independent* if knowing that one event has occurred does not change the probability that the other event has occurred. For example, consider the following two events:

F: The event that a randomly selected student is female

S: The event that a randomly selected student participates in an after-school athletics program

Events F and S would be independent if the probability that a randomly selected student participates in an after-school athletics program is equal to the probability that a randomly selected student who is female participates in an after-school athletics program. If this was the case, knowing that a randomly selected student is female does not change the probability that the selected student participates in an after-school athletics program. Then, F and S would be independent.

5. Based on the definition of independence, are the events "randomly selected student is female" and "randomly selected student participates in an after-school athletics program" independent? Explain.

6. A randomly selected student participates in an after-school athletics program.

    a. What is the probability this student is female?

    b. Using only your answer from part (a), what is the probability that this student is male? Explain how you arrived at your answer.

Exercises 7–11

Consider the data below.

	No Household Member Smokes	At Least One Household Member Smokes	Total
Student Has Asthma	69	113	182
Student Does Not Have Asthma	473	282	755
Total	542	395	937

7. You are asked to determine if the two events "a randomly selected student has asthma" and "a randomly selected student has a household member who smokes" are independent. What probabilities could you calculate to answer this question?

8. Calculate the probabilities you described in Exercise 7.

9. Based on the probabilities you calculated in Exercise 8, are these two events independent or not independent? Explain.

10. Is the probability that a randomly selected student who has asthma and who has a household member who smokes the same as or different from the probability that a randomly selected student who does not have asthma but does have a household member who smokes? Explain your answer.

11. A student is selected at random. The selected student indicates that he has a household member who smokes. What is the probability that the selected student has asthma?

Lesson Summary

Data organized in a two-way frequency table can be used to calculate conditional probabilities.

Two events are independent if knowing that one event has occurred does not change the probability that the second event has occurred.

Probabilities calculated from two-way frequency tables can be used to determine if two events are independent or not independent.

Problem Set

1. Consider the following questions:

   a. A survey of the students at a Midwest high school asked the following questions:

      "Do you use a computer at least 3 times a week to complete your schoolwork?"

      "Are you taking a mathematics class?"

      Do you think the events "a randomly selected student uses a computer at least 3 times a week" and "a randomly selected student is taking a mathematics class" are independent or not independent? Explain your reasoning.

   b. The same survey also asked students the following:

      "Do you participate in any extracurricular activities at your school?"

      "Do you know what you want to do after high school?"

      Do you think the events "a randomly selected student participates in extracurricular activities" and "a randomly selected student knows what she wants to do after completing high school" are independent or not independent? Explain your reasoning.

   c. People attending a professional football game in 2013 completed a survey that included the following questions:

      "Is this the first time you have attended a professional football game?"

      "Do you think football is too violent?"

      Do you think the events "a randomly selected person who completed the survey is attending a professional football game for the first time" and "a randomly selected person who completed the survey thinks football is too violent" are independent or not independent? Explain your reasoning.

2. Complete the table below in a way that would indicate the two events "uses a computer" and "is taking a mathematics class" are independent.

	Uses a Computer at Least 3 Times a Week for Schoolwork	Does Not Use a Computer at Least 3 Times a Week for Schoolwork	Total
In a Mathematics Class			700
Not in a Mathematics Class			
Total	600		1,000

EUREKA MATH™

3. Complete the following hypothetical 1000 table. Are the events "participates in extracurricular activities" and "know what I want to do after high school" independent or not independent? Justify your answer.

	Participates in Extracurricular Activities	Does Not Participate in Extracurricular Activities	Total
Know What I Want to Do After High School			800
Do Not Know What I Want to Do After High School	50		
Total	600		1,000

4. The following hypothetical 1000 table is from Lesson 2:

	No Household Member Smokes	At Least One Household Member Smokes	Total
Student Has Asthma	73	120	193
Student Does Not Have Asthma	506	301	807
Total	579	421	1,000

The actual data from the entire population are given in the table below.

	No Household Member Smokes	At Least One Household Member Smokes	Total
Student Has Asthma	69	113	182
Student Does Not Have Asthma	473	282	755
Total	542	395	937

a. Based on the hypothetical 1000 table, what is the probability that a randomly selected student who has asthma has at least one household member who smokes?

b. Based on the actual data, what is the probability that a randomly selected student who has asthma has at least one household member who smokes (round your answer to 3 decimal places)?

c. Based on the hypothetical 1000 table, what is the probability that a randomly selected student who has no household member who smokes has asthma?

d. Based on the actual data, what is the probability that a randomly selected student who has no household member who smokes has asthma?

e. What do you notice about the probabilities calculated from the actual data and the probabilities calculated from the hypothetical 1000 table?

5. As part of the asthma research, the investigators wondered if students who have asthma are less likely to have a pet at home than students who do not have asthma. They asked the following two questions:

"Do you have asthma?"

"Do you have a pet at home?"

Based on the responses to these questions, you would like to set up a two-way table that you could use to determine if the following two events are independent or not independent:

Event 1: A randomly selected student has asthma.

Event 2: A randomly selected student has a pet at home.

a. How would you label the rows of the two-way table?

b. How would you label the columns of the two-way table?

c. What probabilities would you calculate to determine if Event 1 and Event 2 are independent?

Lesson 4: Calculating Conditional Probabilities and Evaluating Independence Using Two-Way Tables

Lesson 5: Events and Venn Diagrams

Classwork

Example 1: Shading Regions of a Venn Diagram

At a high school, some students play soccer, and some do not. Also, some students play basketball, and some do not. This scenario can be represented by a Venn diagram, as shown below. The circle labeled S represents the students who play soccer, the circle labeled B represents the students who play basketball, and the rectangle represents all the students at the school.

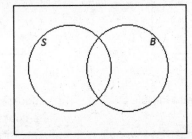

On the Venn diagrams provided, shade the region representing the following instances:

a. The students who play soccer

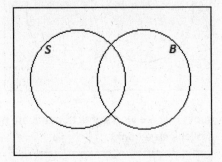

b. The students who do not play soccer

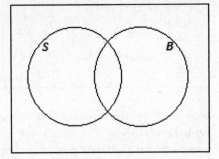

c. The students who play soccer and basketball

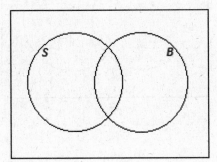

d. The students who play soccer or basketball

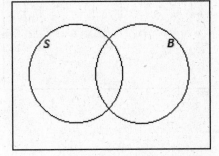

Exercise 1

An online bookstore offers a large selection of books. Some of the books are works of fiction, and some are not. Also, some of the books are available as e-books, and some are not. Let F be the set of books that are works of fiction, and let E be the set of books that are available as e-books. On the Venn diagrams provided, shade the regions representing the following instances:

a. Books that are available as e-books

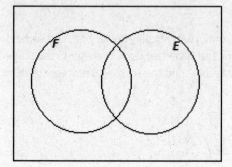

b. Books that are not works of fiction

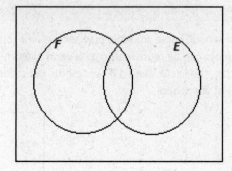

c. Books that are works of fiction and available as e-books

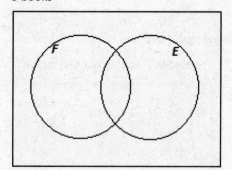

d. Books that are works of fiction or available as e-books

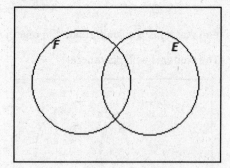

e. Books that are neither works of fiction nor available as e-books

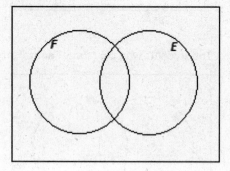

f. Books that are works of fiction that are not available as e-books

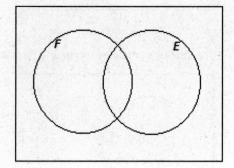

Example 2: Showing Numbers of Possible Outcomes (and Probabilities) in a Venn Diagram

Think again about the school introduced in Example 1. Suppose that 230 students play soccer, 190 students play basketball, and 60 students play both sports. There are a total of 500 students at the school.

    a.   Complete the Venn diagram below by writing the numbers of students in the various regions of the diagram.

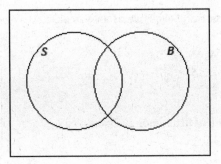

    b.   How many students play basketball but not soccer?

    c.   Suppose that a student will be selected at random from the school.

        i.   What is the probability that the selected student plays both sports?

        ii.   Complete the Venn diagram below by writing the probabilities associated with the various regions of the diagram.

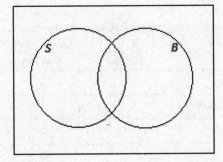

Example 3: Adding and Subtracting Probabilities

Think again about the online bookstore introduced in Exercise 1, and suppose that 62% of the books are works of fiction, 47% are available as e-books, and 14% are available as e-books but are not works of fiction. A book will be selected at random.

a.   Using a Venn diagram, find the following probabilities:

   i.   The book is a work of fiction and available as an e-book.

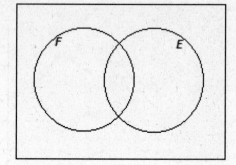

   ii.   The book is neither a work of fiction nor available as an e-book.

b.   Return to the information given at the beginning of the question: 62% of the books are works of fiction, 47% are available as e-books, and 14% are available as e-books but are not works of fiction.

   i.   How would this information be shown in a hypothetical 1000 table? (Show your answers in the table provided below.)

	Fiction	Not Fiction	Total
Available as E-Book			
Not Available as E-Book			
Total			1,000

   ii.   Complete the hypothetical 1000 table given above.

   iii.   Complete the table below showing the probabilities of the events represented by the cells in the table.

	Fiction	Not Fiction	Total
Available as E-Book			
Not Available as E-Book			
Total			

   iv.   How do the probabilities in your table relate to the probabilities you calculated in part (a)?

Exercise 2

When a fish is selected at random from a tank, the probability that it has a green tail is 0.64, the probability that it has red fins is 0.25, and the probability that it has both a green tail and red fins is 0.19.

    a.   Draw a Venn diagram to represent this information.

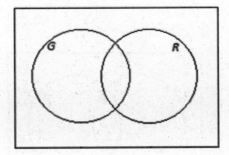

    b.   Find the following probabilities:

        i.   The fish has red fins but does not have a green tail.

        ii.   The fish has a green tail but not red fins.

        iii.   The fish has neither a green tail nor red fins.

    c.   Complete the table below showing the probabilities of the events corresponding to the cells of the table.

	Green Tail	Not Green Tail	Total
Red Fins			
Not Red Fins			
Total			

Exercise 3

In a company, 43% of the employees have access to a fax machine, 38% have access to a fax machine and a scanner, and 24% have access to neither a fax machine nor a scanner. Suppose that an employee will be selected at random. Using a Venn diagram, calculate the probability that the randomly selected employee will not have access to a scanner. (Note that Venn diagrams and probabilities use decimals or fractions, not percentages.) Explain how you used the Venn diagram to determine your answer.

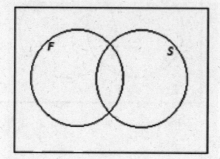

Lesson Summary

In a probability experiment, the events can be represented by circles in a Venn diagram.

Combinations of events using *and*, *or*, and *not* can be shown by shading the appropriate regions of the Venn diagram.

The number of possible outcomes can be shown in each region of the Venn diagram; alternatively, probabilities may be shown. The number of outcomes in a given region (or the probability associated with it) can be calculated by adding or subtracting the known numbers of possible outcomes (or probabilities).

Problem Set

1. On a flight, some of the passengers have frequent-flier status, and some do not. Also, some of the passengers have checked baggage, and some do not. Let the set of passengers who have frequent-flier status be F and the set of passengers who have checked baggage be C. On the Venn diagrams provided, shade the regions representing the following instances:

   a. Passengers who have frequent-flier status and have checked baggage

   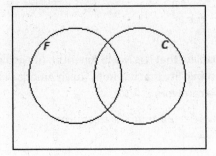

   b. Passengers who have frequent-flier status or have checked baggage

   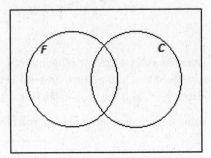

   c. Passengers who do not have both frequent-flier status and checked baggage

   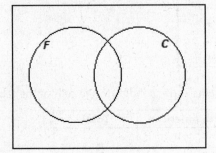

   d. Passengers who have frequent-flier status or do not have checked baggage

   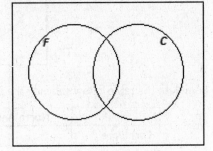

2. For the scenario introduced in Problem 1, suppose that, of the 400 people on the flight, 368 have checked baggage, 228 have checked baggage but do not have frequent-flier status, and 8 have neither frequent-flier status nor checked baggage.

   a. Using a Venn diagram, calculate the following:

      i. The number of people on the flight who have frequent-flier status and have checked baggage

      ii. The number of people on the flight who have frequent-flier status

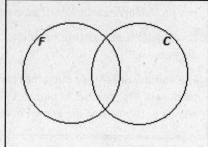

   b. In the Venn diagram provided below, write the probabilities of the events associated with the regions marked with a star (*).

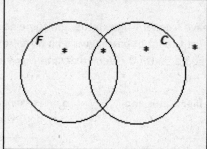

3. When an animal is selected at random from those at a zoo, the probability that it is North American (meaning that its natural habitat is in the North American continent) is 0.65, the probability that it is both North American and a carnivore is 0.16, and the probability that it is neither American nor a carnivore is 0.17.

   a. Using a Venn diagram, calculate the probability that a randomly selected animal is a carnivore.

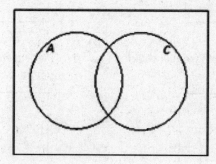

   b. Complete the table below showing the probabilities of the events corresponding to the cells of the table.

	North American	Not North American	Total
Carnivore			
Not Carnivore			
Total			

EUREKA
MATH™

4. This question introduces the mathematical symbols for *and*, *or*, and *not*.

   Considering all the people in the world, let A be the set of Americans (citizens of the United States), and let B be the set of people who have brothers.

   - The set of people who are Americans and have brothers is represented by the shaded region in the Venn diagram below.

   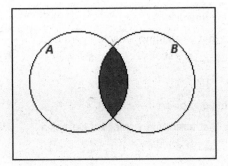

   This set is written $A \cap B$ (read A intersect B), and the probability that a randomly selected person is American and has a brother is written $P(A \cap B)$.

   - The set of people who are Americans or have brothers is represented by the shaded region in the Venn diagram below.

   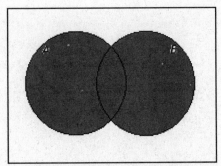

   This set is written $A \cup B$ (read A union B), and the probability that a randomly selected person is American or has a brother is written $P(A \cup B)$.

   - The set of people who are not Americans is represented by the shaded region in the Venn diagram below.

   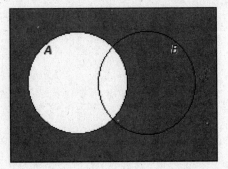

   This set is written A^C (read A complement), and the probability that a randomly selected person is not American is written $P(A^C)$.

©2015 Great Minds. eureka-math.org
ALG II-M4-SE-B2-1.3.0-08.2015

Now, think about the cars available at a dealership. Suppose a car is selected at random from the cars at this dealership. Let the event that the car has manual transmission be denoted by M, and let the event that the car is a sedan be denoted by S. The Venn diagram below shows the probabilities associated with four of the regions of the diagram.

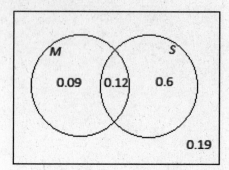

a.   What is the value of $P(M \cap S)$?

b.   Complete this sentence using *and* or *or*:

$P(M \cap S)$ is the probability that a randomly selected car has a manual transmission _____ is a sedan.

c.   What is the value of $P(M \cup S)$?

d.   Complete this sentence using *and* or *or*:

$P(M \cup S)$ is the probability that a randomly selected car has a manual transmission _____ is a sedan.

e.   What is the value of $P(S^C)$?

f.   Explain the meaning of $P(S^C)$.

EUREKA
MATH™

Lesson 6: Probability Rules

Classwork

Example 1: The Complement Rule

In previous lessons, you have seen that to calculate the probability that an event *does not happen,* you can subtract the probability of the event from 1. If the event is denoted by A, then this rule can be written:

$$P(\text{not } A) = 1 - P(A).$$

For example, suppose that the probability that a particular flight is on time is 0.78. What is the probability that the flight is not on time?

Example 2: Formula for Conditional Probability

When a room is randomly selected in a downtown hotel, the probability that the room has a king-sized bed is 0.62, the probability that the room has a view of the town square is 0.43, and the probability that it has a king-sized bed *and* a view of the town square is 0.38. Let A be the event that the room has a king-sized bed, and let B be the event that the room has a view of the town square.

a.   What is the meaning of $P(A \text{ given } B)$ in this context?

b.   Use a hypothetical 1000 table to calculate $P(A \text{ given } B)$.

	A (room has a king-sized bed)	Not A (room does not have a king-sized bed)	Total
B (room has a view of the town square)			
Not B (room does not have a view of the town square)			
Total			

c.   There is also a formula for calculating a conditional probability. The formula for conditional probability is

$$P(A \text{ given } B) = \frac{P(A \text{ and } B)}{P(B)}$$

Use this formula to calculate $P(A \text{ given } B)$, where the events A and B are as defined in this example.

d.   How does the probability you calculated using the formula compare to the probability you calculated using the hypothetical 1000 table?

Exercise 1

A credit card company states that 42% of its customers are classified as long-term cardholders, 35% pay their bills in full each month, and 23% are long-term cardholders who also pay their bills in full each month. Let the event that a randomly selected customer is a long-term cardholder be L and the event that a randomly selected customer pays his bill in full each month be F.

a.   What are the values of $P(L)$, $P(F)$, and $P(L \text{ and } F)$?

b.   Draw a Venn diagram, and label it with the probabilities from part (a).

c.   Use the conditional probability formula to calculate $P(L \text{ given } F)$. (Round your answer to the nearest thousandth.)

©2015 Great Minds. eureka-math.org
ALG II-M4-SE-B2-1.3.0-08.2015

d. Use the conditional probability formula to calculate $P(F$ given $L)$. (Round your answer to the nearest thousandth.)

e. Which is greater, $P(F$ given $L)$ or $P(F)$? Explain why this is relevant.

f. Remember that two events A and B are said to be independent if $P(A$ given $B) = P(A)$. Are the events F and L independent? Explain.

Example 3: Using the Multiplication Rule for Independent Events

A number cube has faces numbered 1 through 6, and a coin has two sides, heads and tails.

The number cube will be rolled, and the coin will be flipped. Find the probability that the cube shows a 4 and the coin lands on heads. Because the events are independent, we can use the multiplication rules we just learned.

If you toss the coin five times, what is the probability you will see a head on all five tosses?

If you tossed the coin five times and got five heads, would you think that this coin is a fair coin? Why or why not?

If you roll the number cube three times, what is the probability that it will show 4 on all three throws?

If you rolled the number cube three times and got a 4 on all three rolls, would you think that this number cube is fair? Why or why not?

Suppose that the credit card company introduced in Exercise 1 states that when a customer is selected at random, the probability that the customer pays his bill in full each month is 0.35, the probability that the customer makes regular online purchases is 0.83, and these two events are independent. What is the probability that a randomly selected customer pays his bill in full each month *and* makes regular online purchases?

Exercise 2

A spinner has a pointer, and when the pointer is spun, the probability that it stops in the red section of the spinner is 0.25.

a. If the pointer is spun twice, what is the probability that it will stop in the red section on both occasions?

b. If the pointer is spun four times, what is the probability that it will stop in the red section on all four occasions? (Round your answer to the nearest thousandth.)

c. If the pointer is spun five times, what is the probability that it never stops on red? (Round your answer to the nearest thousandth.)

Lesson Summary

For any event A, $P(\text{not } A) = 1 - P(A)$.

For any two events A and B, $P(A \text{ given } B) = \dfrac{P(A \text{ and } B)}{P(A)}$.

Events A and B are independent if and only if $P(A \text{ and } B) = P(A)P(B)$.

Problem Set

1. When an avocado is selected at random from those delivered to a food store, the probability that it is ripe is 0.12, the probability that it is bruised is 0.054, and the probability that it is ripe and bruised is 0.019.

   a. Rounding your answers to the nearest thousandth where necessary, find the probability that an avocado randomly selected from those delivered to the store is

      i. Not bruised.

      ii. Ripe given that it is bruised.

      iii. Bruised given that it is ripe.

   b. Which is larger, the probability that a randomly selected avocado is bruised given that it is ripe or the probability that a randomly selected avocado is bruised? Explain in words what this tells you.

   c. Are the events "ripe" and "bruised" independent? Explain.

2. Return to the probability information given in Problem 1. Complete the hypothetical 1000 table given below, and use it to find the probability that a randomly selected avocado is bruised given that it is not ripe. (Round your answer to the nearest thousandth.)

	Ripe	Not Ripe	Total
Bruised			
Not Bruised			
Total			

3. According to the U.S. census website (www.census.gov), based on the U.S. population in 2010, the probability that a randomly selected man is 65 or older is 0.114, and the probability that a randomly selected woman is 65 or older is 0.146. In the questions that follow, round your answers to the nearest thousandth:

   a. If a man is selected at random and a woman is selected at random, what is the probability that both people selected are 65 or older? (Hint: Use the multiplication rule for independent events.)

   b. If two men are selected at random, what is the probability that both of them are 65 or older?

   c. If two women are selected at random, what is the probability that neither of them is 65 or older?

---

4. In a large community, 72% of the people are adults, 78% of the people have traveled outside the state, and 11% are adults who have not traveled outside the state.

   a. Using a Venn diagram or a hypothetical 1000 table, calculate the probability that a randomly selected person from the community is an adult and has traveled outside the state.

   b. Use the multiplication rule for independent events to decide whether the events "is an adult" and "has traveled outside the state" are independent.

5. In a particular calendar year, 10% of the registered voters in a small city are called for jury duty. In this city, people are selected for jury duty at random from all registered voters in the city, and the same individual cannot be called more than once during the calendar year.

   a. What is the probability that a registered voter is not called for jury duty during a particular year?

   b. What is the probability that a registered voter is called for jury duty two years in a row?

6. A survey of registered voters in a city in New York was carried out to assess support for a new school tax. 51% of the respondents supported the school tax. Of those with school-age children, 56% supported the school tax, while only 45% of those who did not have school-age children supported the school tax.

   a. If a person who responded to this survey is selected at random, what is the probability that

      i. The person selected supports the school tax?

      ii. The person supports the school tax given that she does not have school-age children?

   b. Are the two events "has school-age children" and "supports the school tax" independent? Explain how you know this.

   c. Suppose that 35% of those responding to the survey were over the age of 65 and that 10% of those responding to the survey were both over age 65 and supported the school tax. What is the probability that a randomly selected person who responded to this survey supported the school tax given that she was over age 65?

Lesson 7: Probability Rules

Classwork

Exercise 1

When a car is brought to a repair shop for a service, the probability that it will need the transmission fluid replaced is 0.38, the probability that it will need the brake pads replaced is 0.28, and the probability that it will need both the transmission fluid and the brake pads replaced is 0.16. Let the event that a car needs the transmission fluid replaced be T and the event that a car needs the brake pads replaced be B.

    a. What are the values of the following probabilities?

       i. $P(T)$

       ii. $P(B)$

       iii. $P(T \text{ and } B)$

    b. Use the addition rule to find the probability that a randomly selected car needs the transmission fluid or the brake pads replaced.

Exercise 2

Josie will soon be taking exams in math and Spanish. She estimates that the probability she passes the math exam is 0.9, and the probability that she passes the Spanish exam is 0.8. She is also willing to assume that the results of the two exams are independent of each other.

    a. Using Josie's assumption of independence, calculate the probability that she passes both exams.

b.   Find the probability that Josie passes at least one of the exams. (Hint: Passing at least one of the exams is passing math *or* passing Spanish.)

Example 1: Use of the Addition Rule for Disjoint Events

A set of 40 cards consists of the following:

- 10 black cards showing squares
- 10 black cards showing circles
- 10 red cards showing X's
- 10 red cards showing diamonds

A card will be selected at random from the set. Find the probability that the card is black or shows a diamond.

Example 2: Combining Use of the Multiplication and Addition Rules

A red cube has faces labeled 1 through 6, and a blue cube has faces labeled in the same way. The two cubes are rolled. Find the probability of each event.

a.   Both cubes show 6's.

b.   The total score is at least 11.

Exercise 3

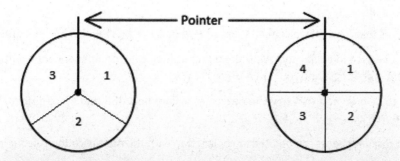

The diagram above shows two spinners. For the first spinner, the scores 1, 2, and 3 are equally likely, and for the second spinner, the scores 1, 2, 3, and 4 are equally likely. Both pointers will be spun. Writing your answers as fractions in lowest terms, find the probability of each event.

a. The total of the scores on the two spinners is 2.

b. The total of the scores on the two spinners is 3.

c. The total of the scores on the two spinners is 5.

d. The total of the scores on the two spinners is not 5.

Lesson Summary

The addition rule states that for any two events A and B, $P(A \text{ or } B) = P(A) + P(B) - P(A \text{ and } B)$.

The addition rule can be used in conjunction with the multiplication rule for independent events: Events A and B are independent if and only if $P(A \text{ and } B) = P(A)P(B)$.

Two events are said to be *disjoint* if they have no outcomes in common. If A and B are disjoint events, then $P(A \text{ or } B) = P(A) + P(B)$.

The addition rule for disjoint events can be used in conjunction with the multiplication rule for independent events.

Problem Set

1. Of the works of art at a large gallery, 59% are paintings, and 83% are for sale. When a work of art is selected at random, let the event that it is a painting be A and the event that it is for sale be B.

   a. What are the values of $P(A)$ and $P(B)$?

   b. Suppose you are told that $P(A \text{ and } B) = 0.51$. Find $P(A \text{ or } B)$.

   c. Suppose now that you are not given the information in part (b), but you are told that the events A and B are independent. Find $P(A \text{ or } B)$.

2. A traveler estimates that, for an upcoming trip, the probability of catching malaria is 0.18, the probability of catching typhoid is 0.13, and the probability of catching neither of the two diseases is 0.75.

   a. Draw a Venn diagram to represent this information.

   b. Calculate the probability of catching both of the diseases.

   c. Are the events "catches malaria" and "catches typhoid" independent? Explain your answer.

3. A deck of 40 cards consists of the following:
   - 10 black cards showing squares, numbered 1–10
   - 10 black cards showing circles, numbered 1–10
   - 10 red cards showing X's, numbered 1–10
   - 10 red cards showing diamonds, numbered 1–10

   A card will be selected at random from the deck.

   a. i. Are the events "the card shows a square" and "the card is red" disjoint? Explain.

      ii. Calculate the probability that the card will show a square or will be red.

   b. i. Are the events "the card shows a 5" and "the card is red" disjoint? Explain.

      ii. Calculate the probability that the card will show a 5 or will be red.

4. The diagram below shows a spinner. When the pointer is spun, it is equally likely to stop on 1, 2, or 3. The pointer will be spun three times. Expressing your answers as fractions in lowest terms, find the probability, and explain how the answer was determined that the total of the values from all thee spins is

   a. 9.

   b. 8.

   c. 7.

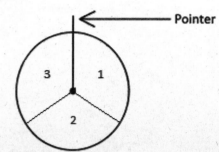

5. A number cube has faces numbered 1 through 6, and a coin has two sides, heads and tails. The number cube will be rolled once, and the coin will be flipped once. Find the probabilities of the following events. (Express your answers as fractions in lowest terms.)

   a. The number cube shows a 6.

   b. The coin shows heads.

   c. The number cube shows a 6, and the coin shows heads.

   d. The number cube shows a 6, or the coin shows heads.

6. Kevin will soon be taking exams in math, physics, and French. He estimates the probabilities of his passing these exams to be as follows:

   ▪ Math: 0.9

   ▪ Physics: 0.8

   ▪ French: 0.7

   Kevin is willing to assume that the results of the three exams are independent of each other. Find the probability of each event.

   a. Kevin will pass all three exams.

   b. Kevin will pass math but fail the other two exams.

   c. Kevin will pass exactly one of the three exams.

This page intentionally left blank

Lesson 8: Distributions—Center, Shape, and Spread

Classwork

Example 1: Center, Shape, and Spread

Have you ever noticed how sometimes batteries seem to last a long time, and other times the batteries seem to last only a short time?

The histogram below shows the distribution of battery life (hours) for a sample of 40 batteries of the same brand. When studying a distribution, it is important to think about the shape, center, and spread of the data.

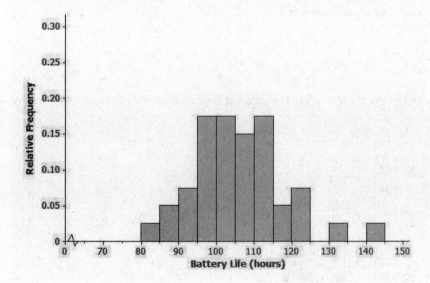

Exercises 1–9

1. Would you describe the distribution of battery life as approximately symmetric or as skewed? Explain your answer.

2. Is the mean of the battery life distribution closer to 95, 105, or 115 hours? Explain your answer.

EUREKA
MATH™

Lesson 8:   Distributions—Center, Shape, and Spread

S.59

©2015 Great Minds. eureka-math.org
ALG II-M4-SE-B2-1.3.0-08.2015

3. Consider 5, 10, or 25 hours as an estimate of the standard deviation for the battery life distribution.

    a. Consider 5 hours as an estimate of the standard deviation. Is it a reasonable description of a typical distance from the mean? Explain your answer.

    b. Consider 10 hours as an estimate of the standard deviation. Is it a reasonable description of a typical distance from the mean? Explain your answer.

    c. Consider 25 hours as an estimate of the standard deviation. Is it a reasonable description of a typical distance from the mean? Explain your answer.

The histogram below shows the distribution of the greatest drop (in feet) for 55 major roller coasters in the United States.

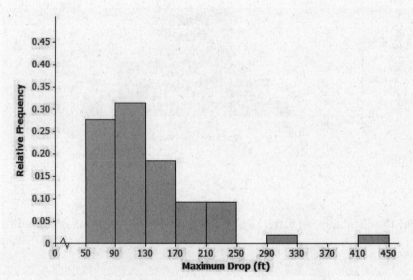

4. Would you describe this distribution of roller coaster maximum drop as approximately symmetric or as skewed? Explain your answer.

5. Is the mean of the maximum drop distribution closer to 90, 135, or 240 feet? Explain your answer.

6. Is the standard deviation of the maximum drop distribution closer to 40, 70, or 100 hours? Explain your answer.

7. Consider the following histograms: Histogram 1, Histogram 2, Histogram 3, and Histogram 4. Descriptions of four distributions are also given. Match the description of a distribution with the appropriate histogram.

Histogram	Distribution
1	
2	
3	
4	

Description of distributions:

Distribution	Shape	Mean	Standard Deviation
A	Skewed to the right	100	10
B	Approximately symmetric, mound shaped	100	10
C	Approximately symmetric, mound shaped	100	40
D	Skewed to the right	100	40

Histograms:

Histogram 1

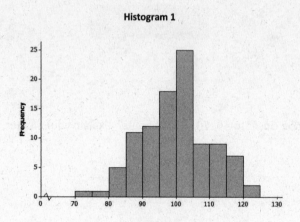

Histogram 2

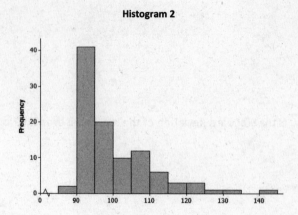

Histogram 3

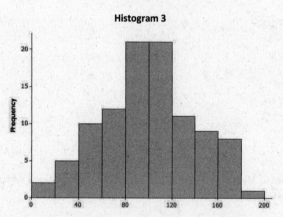

Histogram 4

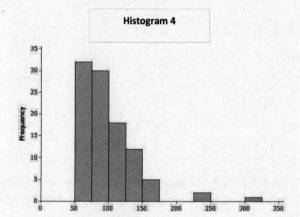

Lesson 8: Distributions—Center, Shape, and Spread

EUREKA MATH™

8. The histogram below shows the distribution of gasoline tax per gallon for the 50 states and the District of Columbia in 2010. Describe the shape, center, and spread of this distribution.

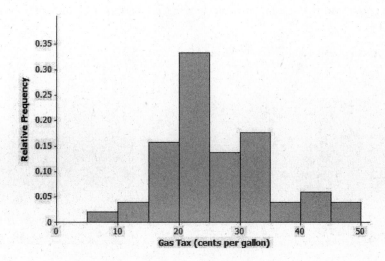

9. The histogram below shows the distribution of the number of automobile accidents per year for every 1,000 people in different occupations. Describe the shape, center, and spread of this distribution.

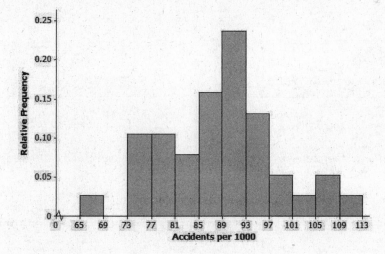

Lesson Summary

Distributions are described by the shape (symmetric or skewed), the center, and the spread (variability) of the distribution.

A distribution that is approximately symmetric can take different forms.

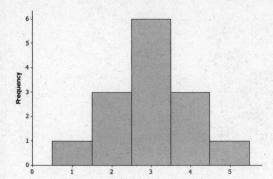
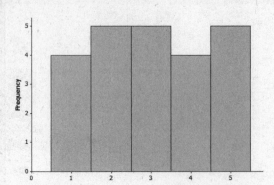

A distribution is described as *mound shaped* if it is approximately symmetric and has a single peak.

A distribution is *skewed to the right* or *skewed to the left* if one of its tails is longer than the other.

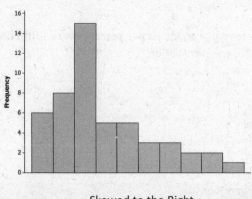

Skewed to the Right

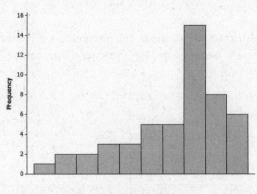

Skewed to the Left

The *mean of a distribution* is interpreted as a typical value and is the average of the data values that make up the distribution.

The *standard deviation* is a value that describes a typical distance from the mean.

EUREKA
MATH™

Problem Set

1. For each of the following histograms, describe the shape, and give estimates of the mean and standard deviation of the distributions:

    a. Distribution of head circumferences (mm)

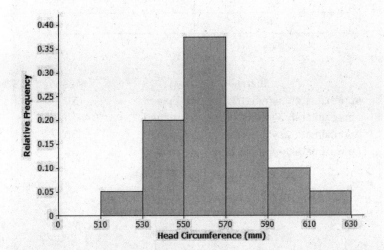

    b. Distribution of NBA arena seating capacity

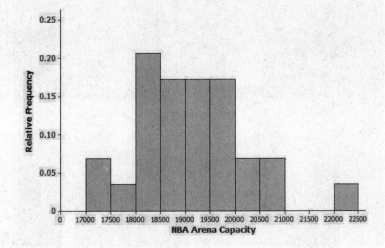

2. For the each of the following, match the description of each distribution with the appropriate histogram:

Histogram	Distribution
1	
2	
3	
4	

Description of distributions:

Distribution	Shape	Mean	Standard Deviation
A	Approximately symmetric, mound shaped	50	5
B	Approximately symmetric, mound shaped	50	10
C	Approximately symmetric, mound shaped	30	10
D	Approximately symmetric, mound shaped	30	5

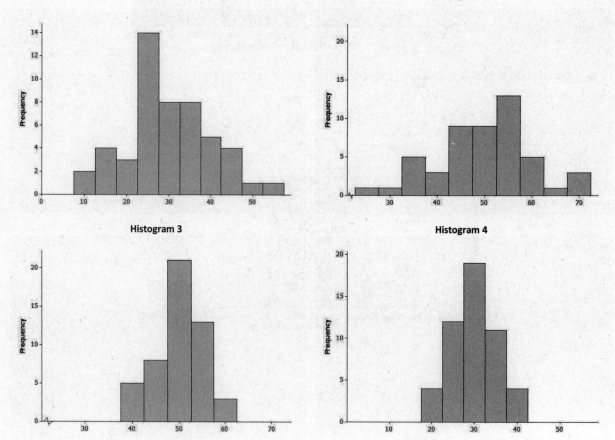

Lesson 8:   Distributions—Center, Shape, and Spread

EUREKA MATH™

3. Following are the number of calories in a basic hamburger (one meat patty with no cheese) at various fast food restaurants around the country:

380, 790, 680, 460, 725, 1130, 240, 260, 930, 331, 710, 680, 1080, 612, 1180, 400, 866, 700, 1060, 270, 550, 380, 940, 280, 940, 550, 549, 937, 820, 870, 250, 740

a. Draw a dot plot on the scale below.

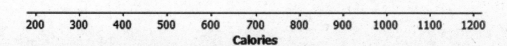

b. Describe the shape of the calorie distribution.

c. Using technology, find the mean and standard deviation of the calorie data.

d. Why do you think there is a lot of variability in the calorie data?

This page intentionally left blank

Lesson 9: Using a Curve to Model a Data Distribution

Classwork

Example 1: Heights of Dinosaurs and the Normal Curve

A paleontologist studies prehistoric life and sometimes works with dinosaur fossils. The table below shows the distribution of heights (rounded to the nearest inch) of 660 procompsognathids, otherwise known as compys.

The heights were determined by studying the fossil remains of the compys.

Height (cm)	Number of Compys	Relative Frequency
26	1	0.002
27	5	0.008
28	12	0.018
29	22	0.033
30	40	0.061
31	60	0.091
32	90	0.136
33	100	0.152
34	100	0.152
35	90	0.136
36	60	0.091
37	40	0.061
38	22	0.033
39	12	0.018
40	5	0.008
41	1	0.002
Total	660	1.000

Exercises 1–8

The following is a relative frequency histogram of the compy heights:

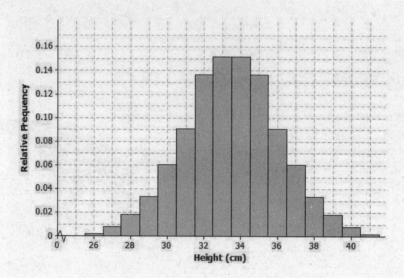

1. What does the relative frequency of 0.136 mean for the height of 32 cm?

2. What is the width of each bar? What does the height of the bar represent?

3. What is the area of the bar that represents the relative frequency for compys with a height of 32 cm?

4. The mean of the distribution of compy heights is 33.5 cm, and the standard deviation is 2.56 cm. Interpret the mean and standard deviation in this context.

5. Mark the mean on the graph, and mark *one* deviation above and below the mean.

   a. Approximately what percent of the values in this data set are within one standard deviation of the mean (i.e., between 33.5 cm $- 2.56$ cm $= 30.94$ cm and 33.5 cm $+ 2.56$ cm $= 36.06$ cm)?

   b. Approximately what percent of the values in this data set are within *two* standard deviations of the mean?

6. Draw a smooth curve that comes reasonably close to passing through the midpoints of the tops of the bars in the histogram. Describe the shape of the distribution.

7. Shade the area of the histogram that represents the proportion of heights that are within one standard deviation of the mean.

8. Based on our analysis, how would you answer the question, "How tall was a compy?"

Example 2: Gas Mileage and the Normal Distribution

A normal curve is a smooth curve that is symmetric and bell shaped. Data distributions that are mound shaped are often modeled using a normal curve, and we say that such a distribution is approximately normal. One example of a distribution that is approximately normal is the distribution of compy heights from Example 1. Distributions that are approximately normal occur in many different settings. For example, a salesman kept track of the gas mileage for his car over a 25-week span.

The mileages (miles per gallon rounded to the nearest whole number) were

23, 27, 27, 28, 25, 26, 25, 29, 26, 27, 24, 26, 26, 24, 27, 25, 28, 25, 26, 25, 29, 26, 27, 24, 26.

Exercise 9

9. Consider the following:

    a. Use technology to find the mean and standard deviation of the mileage data. How did you use technology to assist you?

    b. Calculate the relative frequency of each of the mileage values. For example, the mileage of 26 mpg has a frequency of 7. To find the relative frequency, divide 7 by 25, the total number of mileages recorded. Complete the following table:

Mileage (mpg)	Frequency	Relative Frequency
23		
24		
25		
26	7	
27		
28		
29		
Total	25	

c. Construct a relative frequency histogram using the scale below.

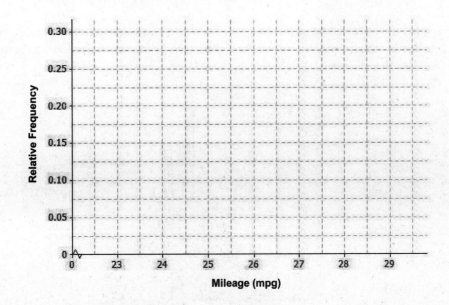

d. Describe the shape of the mileage distribution. Draw a smooth curve that comes reasonably close to passing through the midpoints of the tops of the bars in the histogram. Is this approximately a normal curve?

e. Mark the mean on the histogram. Mark one standard deviation to the left and right of the mean. Shade the area of the histogram that represents the proportion of mileages that are within one standard deviation of the mean. Find the proportion of the data within one standard deviation of the mean.

Lesson Summary

- A normal curve is symmetric and bell shaped. The mean of a normal distribution is located in the center of the distribution. Areas under a normal curve can be used to estimate the proportion of the data values that fall within a given interval.

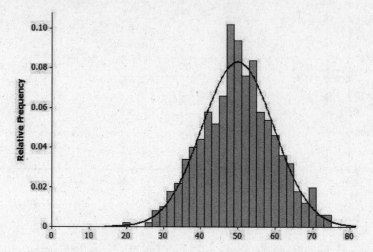

- When a distribution is skewed, it is not appropriate to model the data distribution with a normal curve.

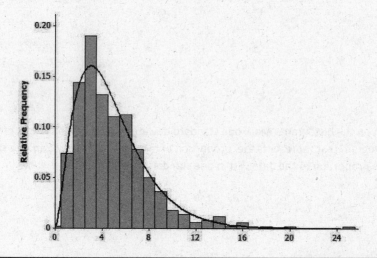

EUREKA
MATH™

Problem Set

1. Periodically the U.S. Mint checks the weight of newly minted nickels. Below is a histogram of the weights (in grams) of a random sample of 100 new nickels.

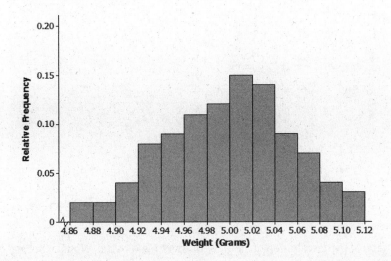

a. The mean and standard deviation of the distribution of nickel weights are 5.00 grams and 0.06 gram, respectively. Mark the mean on the histogram. Mark one standard deviation above the mean and one standard deviation below the mean.

b. Describe the shape of the distribution. Draw a smooth curve that comes reasonably close to passing through the midpoints of the tops of the bars in the histogram. Is this approximately a normal curve?

c. Shade the area of the histogram that represents the proportion of weights that are within one standard deviation of the mean. Find the proportion of the data within one standard deviation of the mean.

2. Below is a relative frequency histogram of the gross (in millions of dollars) for the all-time top-grossing American movies (as of the end of 2012). Gross is the total amount of money made before subtracting out expenses, like advertising costs and actors' salaries.

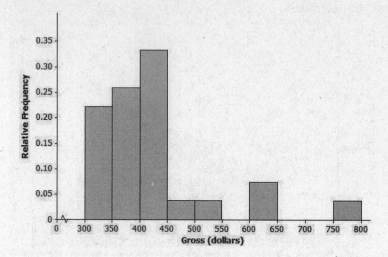

a. Describe the shape of the distribution of all-time top-grossing movies. Would a normal curve be the best curve to model this distribution? Explain your answer.

b. Which of the following is a reasonable estimate for the mean of the distribution? Explain your choice.

 i. 325 million

 ii. 375 million

 iii. 425 million

c. Which of the following is a reasonable estimate for the sample standard deviation? Explain your choice.

 i. 50 million

 ii. 100 million

 iii. 200 million

3. Below is a histogram of the top speed of different types of animals.

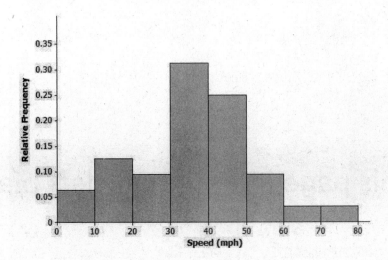

a. Describe the shape of the top speed distribution.

b. Estimate the mean and standard deviation of this distribution. Describe how you made your estimate.

c. Draw a smooth curve that is approximately a normal curve. The actual mean and standard deviation of this data set are 34.1 mph and 15.3 mph, respectively. Shade the area of the histogram that represents the proportion of speeds that are within one standard deviation of the mean.

This page intentionally left blank

Lesson 10: Normal Distributions

Classwork

Exercise 1

Consider the following data distributions. In the previous lesson, you distinguished between distributions that were approximately normal and those that were not. For each of the following distributions, indicate if it is approximately normal, skewed, or neither, and explain your choice:

a.

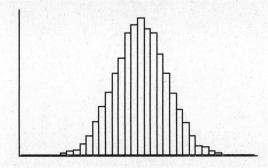

b.

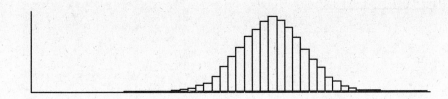

c.

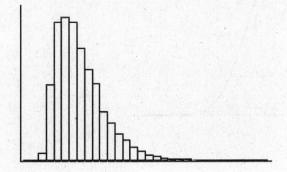

d.

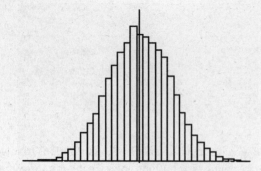

e.

A normal distribution is a distribution that has a particular symmetric mound shape, as shown below.

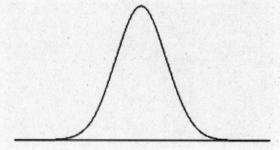

Lesson 10:   Normal Distributions

EUREKA
MATH™

Exercise 2

When calculating probabilities associated with normal distributions, z-scores are used. A z-score for a particular value measures the number of standard deviations away from the mean. A positive z-score corresponds to a value that is above the mean, and a negative z-score corresponds to a value that is below the mean. The letter z is used to represent a variable that has a standard normal distribution where the mean is 0 and standard deviation is 1. This distribution was used to define a z-score. A z-score is calculated by

$$z = \frac{\text{value} - \text{mean}}{\text{standard deviation}}.$$

a. The prices of the printers in a store have a mean of \$240 and a standard deviation of \$50. The printer that you eventually choose costs \$340.

   i. What is the z-score for the price of your printer?

   ii. How many standard deviations above the mean was the price of your printer?

b. Ashish's height is 63 inches. The mean height for boys at his school is 68.1 inches, and the standard deviation of the boys' heights is 2.8 inches.

   i. What is the z-score for Ashish's height? (Round your answer to the nearest hundredth.)

   ii. What is the meaning of this value?

c. Explain how a z-score is useful in describing data.

Example 1: Use of z-Scores and a Graphing Calculator to Find Normal Probabilities

A swimmer named Amy specializes in the 50-meter backstroke. In competition, her mean time for the event is 39.7 seconds, and the standard deviation of her times is 2.3 seconds. Assume that Amy's times are approximately normally distributed.

a. Estimate the probability that Amy's time is between 37 and 44 seconds.

b. Using z-scores and a graphing calculator and rounding your answers to the nearest thousandth, find the probability that Amy's time in her next race is between 37 and 44 seconds.

c. Estimate the probability that Amy's time is more than 45 seconds.

d. Using z-scores and a graphing calculator and rounding your answers to the nearest thousandth, find the probability that Amy's time in her next race is more than 45 seconds.

e.   What is the probability that Amy's time would be at least 45 seconds?

f.   Using z-scores and a graphing calculator and rounding your answers to the nearest thousandth, find the probability that Amy's time in her next race is less than 36 seconds.

Exercise 3

The distribution of lifetimes of a particular brand of car tires has a mean of 51,200 miles and a standard deviation of 8,200 miles.

a.   Assuming that the distribution of lifetimes is approximately normally distributed and rounding your answers to the nearest thousandth, find the probability of each event.

   i.   A randomly selected tire lasts between 55,000 and 65,000 miles.

   ii.   A randomly selected tire lasts less than 48,000 miles.

iii.   A randomly selected tire lasts at least 41,000 miles.

b.   Explain the meaning of the probability that you found in part (a)(iii).

Exercise 4

Think again about the brand of tires described in Exercise 3. What is the probability that the lifetime of a randomly selected tire is within 10,000 miles of the mean lifetime for tires of this brand?

Example 2: Using Table of Standard Normal Curve Areas

The standard normal distribution is the normal distribution with a mean of 0 and a standard deviation of 1. The diagrams below show standard normal distribution curves. Use a table of standard normal curve areas to determine the shaded areas.

a.

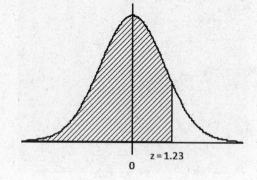

b.

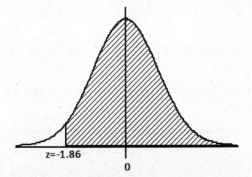

c.

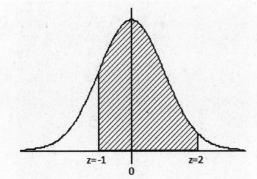

Lesson Summary

A *normal distribution* is a continuous distribution that has the particular symmetric mound-shaped curve that is shown at the beginning of the lesson.

Probabilities associated with normal distributions are determined using z-scores and can be found using a graphing calculator or tables of standard normal curve areas.

Problem Set

1. Which of the following histograms show distributions that are approximately normal?

a.

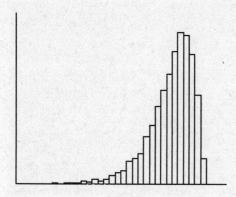

b.

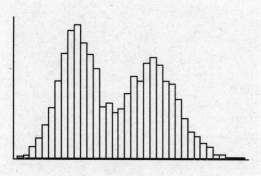

c.

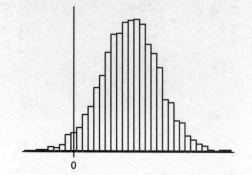

2. Suppose that a particular medical procedure has a cost that is approximately normally distributed with a mean of $19,800 and a standard deviation of $2,900. For a randomly selected patient, find the probabilities of the following events. (Round your answers to the nearest thousandth.)

    a. The procedure costs between $18,000 and $22,000.

    b. The procedure costs less than $15,000.

    c. The procedure costs more than $17,250.

3. Consider the medical procedure described in the previous question, and suppose a patient is charged $24,900 for the procedure. The patient is reported as saying, "I've been charged an outrageous amount!" How justified is this comment? Use probability to support your answer.

4. Think again about the medical procedure described in Problem 2.

    a. Rounding your answers to the nearest thousandth, find the probability of each instance for a randomly selected patient.

        i.   The cost of the procedure is within two standard deviations of the mean cost.

        ii. The cost of the procedure is more than one standard deviation from the mean cost.

    b. If the mean or the standard deviation were to be changed, would your answers to part (a) be affected? Explain.

5. Use a table of standard normal curve areas to find the following:

    a. The area to the left of $z = 0.56$

    b. The area to the right of $z = 1.20$

    c. The area to the left of $z = -1.47$

    d. The area to the right of $z = -0.35$

    e. The area between $z = -1.39$ and $z = 0.80$

    f. Choose a response from parts (a) through (f), and explain how you determined your answer.

This page intentionally left blank

Standard Normal Curve Areas

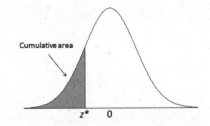

Cumulative area

z^* 0

Z	0.00	0.01	0.02	0.03	0.04	0.05	0.06	0.07	0.08	0.09
−3.8	0.0001	0.0001	0.0001	0.0001	0.0001	0.0001	0.0001	0.0001	0.0001	0.0001
−3.7	0.0001	0.0001	0.0001	0.0001	0.0001	0.0001	0.0001	0.0001	0.0001	0.0001
−3.6	0.0002	0.0002	0.0001	0.0001	0.0001	0.0001	0.0001	0.0001	0.0001	0.0001
−3.5	0.0002	0.0002	0.0002	0.0002	0.0002	0.0002	0.0002	0.0002	0.0002	0.0002
−3.4	0.0003	0.0003	0.0003	0.0003	0.0003	0.0003	0.0003	0.0003	0.0003	0.0002
−3.3	0.0005	0.0005	0.0005	0.0004	0.0004	0.0004	0.0004	0.0004	0.0004	0.0003
−3.2	0.0007	0.0007	0.0006	0.0006	0.0006	0.0006	0.0006	0.0005	0.0005	0.0005
−3.1	0.0010	0.0009	0.0009	0.0009	0.0008	0.0008	0.0008	0.0008	0.0007	0.0007
−3.0	0.0013	0.0013	0.0013	0.0012	0.0012	0.0011	0.0011	0.0011	0.0010	0.0010
−2.9	0.0019	0.0018	0.0018	0.0017	0.0016	0.0016	0.0015	0.0015	0.0014	0.0014
−2.8	0.0026	0.0025	0.0024	0.0023	0.0023	0.0022	0.0021	0.0021	0.0020	0.0019
−2.7	0.0035	0.0034	0.0033	0.0032	0.0031	0.0030	0.0029	0.0028	0.0027	0.0026
−2.6	0.0047	0.0045	0.0044	0.0043	0.0041	0.0040	0.0039	0.0038	0.0037	0.0036
−2.5	0.0062	0.0060	0.0059	0.0057	0.0055	0.0054	0.0052	0.0051	0.0049	0.0048
−2.4	0.0082	0.0080	0.0078	0.0075	0.0073	0.0071	0.0069	0.0068	0.0066	0.0064
−2.3	0.0107	0.0104	0.0102	0.0099	0.0096	0.0094	0.0091	0.0089	0.0087	0.0084
−2.2	0.0139	0.0136	0.0132	0.0129	0.0125	0.0122	0.0119	0.0116	0.0113	0.0110
−2.1	0.0179	0.0174	0.0160	0.0166	0.0162	0.0158	0.0154	0.0150	0.0146	0.0143
−2.0	0.0228	0.0222	0.0217	0.0212	0.0207	0.0202	0.0197	0.0192	0.0188	0.0183
−1.9	0.0287	0.0281	0.0274	0.0268	0.0262	0.0256	0.0250	0.0244	0.0239	0.0233
−1.8	0.0359	0.0351	0.0344	0.0336	0.0329	0.0322	0.0314	0.0307	0.0301	0.0294
−1.7	0.0446	0.0436	0.0427	0.0418	0.0409	0.0401	0.0392	0.0384	0.0375	0.0367
−1.6	0.0548	0.0537	0.0526	0.0516	0.0505	0.0495	0.0485	0.0475	0.0465	0.0455
−1.5	0.0668	0.0655	0.0643	0.0630	0.0618	0.0606	0.0594	0.0582	0.0571	0.0599
−1.4	0.0808	0.0793	0.0778	0.0764	0.0749	0.0735	0.0721	0.0708	0.0694	0.0681
−1.3	0.0968	0.0951	0.0934	0.0918	0.0901	0.0885	0.0869	0.0853	0.0838	0.0823
−1.2	0.1151	0.1131	0.1112	0.1093	0.1075	0.1056	0.1038	0.1020	0.1003	0.0985
−1.1	0.1357	0.1335	0.1314	0.1292	0.1271	0.1251	0.1230	0.1210	0.1190	0.1170
−1.0	0.1587	0.1562	0.1539	0.1515	0.1492	0.1469	0.1446	0.1423	0.1401	0.1379
−0.9	0.1841	0.1814	0.1788	0.1762	0.1736	0.1711	0.1685	0.1660	0.1635	0.1611
−0.8	0.2119	0.2090	0.2061	0.2033	0.2005	0.1977	0.1949	0.1922	0.1894	0.1867
−0.7	0.2420	0.2389	0.2358	0.2327	0.2296	0.2266	0.2236	0.2206	0.2177	0.2148
−0.6	0.2743	0.2709	0.2676	0.2643	0.2611	0.2578	0.2546	0.2514	0.2483	0.2451
−0.5	0.3085	0.3050	0.3015	0.2981	0.2946	0.2912	0.2877	0.2843	0.2810	0.2776
−0.4	0.3446	0.3409	0.3372	0.3336	0.3300	0.3264	0.3228	0.3192	0.3156	0.3121
−0.3	0.3821	0.3783	0.3745	0.3707	0.3669	0.3632	0.3594	0.3557	0.3520	0.3483
−0.2	0.4207	0.4168	0.4129	0.4090	0.4052	0.4013	0.3974	0.3936	0.3897	0.3859
−0.1	0.4602	0.4562	0.4522	0.4483	0.4443	0.4404	0.4364	0.4325	0.4286	0.4247
−0.0	0.5000	0.4960	0.4920	0.4880	0.4840	0.4801	0.4761	0.4721	0.4681	0.4641

z	0.00	0.01	0.02	0.03	0.04	0.05	0.06	0.07	0.08	0.09
0.0	0.5000	0.5040	0.5080	0.5120	0.5160	0.5199	0.5239	0.5279	0.5319	0.5359
0.1	0.5398	0.5438	0.5478	0.5517	0.5557	0.5596	0.5636	0.5675	0.5714	0.5753
0.2	0.5793	0.5832	0.5871	0.5910	0.5948	0.5987	0.6026	0.6064	0.6103	0.6141
0.3	0.6179	0.6217	0.6255	0.6293	0.6331	0.6368	0.6406	0.6443	0.6480	0.6517
0.4	0.6554	0.6591	0.6628	0.6664	0.6700	0.6736	0.6772	0.6808	0.6844	0.6879
0.5	0.6915	0.6950	0.6985	0.7019	0.7054	0.7088	0.7123	0.7157	0.7190	0.7224
0.6	0.7257	0.7291	0.7324	0.7357	0.7389	0.7422	0.7454	0.7486	0.7517	0.7549
0.7	0.7580	0.7611	0.7642	0.7673	0.7704	0.7734	0.7764	0.7794	0.7823	0.7852
0.8	0.7881	0.7910	0.7939	0.7967	0.7995	0.8023	0.8051	0.8078	0.8106	0.8133
0.9	0.8159	0.8186	0.8212	0.8238	0.8264	0.8289	0.8315	0.8340	0.8365	0.8389
1.0	0.8413	0.8438	0.8461	0.8485	0.8508	0.8531	0.8554	0.8577	0.8599	0.8621
1.1	0.8643	0.8665	0.8686	0.8708	0.8729	0.8749	0.8770	0.8790	0.8810	0.8830
1.2	0.8849	0.8869	0.8888	0.8907	0.8925	0.8944	0.8962	0.8980	0.8997	0.9015
1.3	0.9032	0.9049	0.9066	0.9082	0.9099	0.9115	0.9131	0.9147	0.9162	0.9177
1.4	0.9192	0.9207	0.9222	0.9236	0.9251	0.9265	0.9279	0.9292	0.9306	0.9319
1.5	0.9332	0.9345	0.9357	0.9370	0.9382	0.9394	0.9406	0.9418	0.9429	0.9441
1.6	0.9452	0.9463	0.9474	0.9484	0.9495	0.9505	0.9515	0.9525	0.9535	0.9545
1.7	0.9554	0.9564	0.9573	0.9582	0.9591	0.9599	0.9608	0.9616	0.9625	0.9633
1.8	0.9641	0.9649	0.9656	0.9664	0.9671	0.9678	0.9686	0.9693	0.9699	0.9706
1.9	0.9713	0.9719	0.9726	0.9732	0.9738	0.9744	0.9750	0.9756	0.9761	0.9767
2.0	0.9772	0.9778	0.9783	0.9788	0.9793	0.9798	0.9803	0.9808	0.9812	0.9817
2.1	0.9821	0.9826	0.9830	0.9834	0.9838	0.9842	0.9846	0.9850	0.9854	0.9857
2.2	0.9861	0.9864	0.9868	0.9871	0.9875	0.9878	0.9881	0.9884	0.9887	0.9890
2.3	0.9893	0.9896	0.9898	0.9901	0.9904	0.9906	0.9909	0.9911	0.9913	0.9916
2.4	0.9918	0.9920	0.9922	0.9925	0.9927	0.9929	0.9931	0.9932	0.9934	0.9936
2.5	0.9938	0.9940	0.9941	0.9943	0.9945	0.9946	0.9948	0.9949	0.9951	0.9952
2.6	0.9953	0.9955	0.9956	0.9957	0.9959	0.9960	0.9961	0.9962	0.9963	0.9964
2.7	0.9965	0.9966	0.9967	0.9968	0.9969	0.9970	0.9971	0.9972	0.9973	0.9974
2.8	0.9974	0.9975	0.9976	0.9977	0.9977	0.9978	0.9979	0.9979	0.9980	0.9981
2.9	0.9981	0.9982	0.9982	0.9983	0.9984	0.9984	0.9985	0.9985	0.9986	0.9986
3.0	0.9987	0.9987	0.9987	0.9988	0.9988	0.9989	0.9989	0.9989	0.9990	0.9990
3.1	0.9990	0.9991	0.9991	0.9991	0.9992	0.9992	0.9992	0.9992	0.9993	0.9993
3.2	0.9993	0.9993	0.9994	0.9994	0.9994	0.9994	0.9994	0.9995	0.9995	0.9995
3.3	0.9995	0.9995	0.9995	0.9996	0.9996	0.9996	0.9996	0.9996	0.9996	0.9997
3.4	0.9997	0.9997	0.9997	0.9997	0.9997	0.9997	0.9997	0.9997	0.9997	0.9998
3.5	0.9998	0.9998	0.9998	0.9998	0.9998	0.9998	0.9998	0.9998	0.9998	0.9998
3.6	0.9998	0.9998	0.9999	0.9999	0.9999	0.9999	0.9999	0.9999	0.9999	0.9999
3.7	0.9999	0.9999	0.9999	0.9999	0.9999	0.9999	0.9999	0.9999	0.9999	0.9999
3.8	0.9999	0.9999	0.9999	0.9999	0.9999	0.9999	0.9999	0.9999	0.9999	0.9999

©2015 Great Minds. eureka-math.org
ALG II-M4-SE-B2-1.3.0-08.2015

Lesson 11: Normal Distributions

Classwork

Example 1: Calculation of Normal Probabilities Using *z*-Scores and Tables of Standard Normal Areas

The U.S. Department of Agriculture (USDA), in its Official Food Plans (www.cnpp.usda.gov), states that the average cost of food for a 14- to 18-year-old male (on the Moderate-Cost Plan) is $261.50 per month. Assume that the monthly food cost for a 14- to 18-year-old male is approximately normally distributed with a mean of $261.50 and a standard deviation of $16.25.

    a. Use a table of standard normal curve areas to find the probability that the monthly food cost for a randomly selected 14- to 18-year-old male is

        i.   Less than $280.

        ii.   More than $270.

        iii. More than $250.

iv.   Between $240 and $275.

b.   Explain the meaning of the probability that you found in part (a)(iv).

Exercise 1

The USDA document described in Example 1 also states that the average cost of food for a 14- to 18-year-old female (again, on the Moderate-Cost Plan) is $215.20 per month. Assume that the monthly food cost for a 14- to 18-year-old female is approximately normally distributed with a mean of $215.20 and a standard deviation of $14.85.

a.   Use a table of standard normal curve areas to find the probability that the monthly food cost for a randomly selected 14- to 18-year-old female is

i.   Less than $225.

ii.   Less than $200.

iii.   More than $250.

iv.   Between $190 and $220.

b.   Explain the meaning of the probability that you found in part (a)(iv).

Example 2: Use of a Graphing Calculator to Find Normal Probabilities Directly

Return to the information given in Example 1. Using a graphing calculator, and *without* using z-scores, find the probability (rounded to the nearest thousandth) that the monthly food cost for a randomly selected 14- to 18-year-old male is

a.   Between $260 and $265.

b.   At least $252.

c.   At most $248.

Exercise 2

Return to the information given in Exercise 1.

a.   In Exercise 1, you calculated the probability that the monthly food cost for a randomly selected 14- to 18-year-old female is between $190 and $220. Would the probability that the monthly food cost for a randomly selected 14- to 18-year-old female is between $195 and $230 be greater than or smaller than the probability for between $190 and $220? Explain your thinking.

b.   Do you think that the probability that the monthly food cost for a randomly selected 14- to 18-year-old female is between $195 and $230 is closer to 0.50, 0.75, or 0.90? Explain your thinking.

c.   Using a graphing calculator, and without using z-scores, find the probability (rounded to the nearest thousandth) that the monthly food cost for a randomly selected 14- to 18-year-old female is between $195 and $230. Is this probability consistent with your answer to part (b)?

d.   How does the probability you calculated in part (c) compare to the probability that would have been obtained using the table of normal curve areas?

e.   What is one advantage to using a graphing calculator to calculate this probability?

f.   In Exercise 1, you calculated the probability that the monthly food cost for a randomly selected 14- to 18-year-old female is at most $200. Would the probability that the monthly food cost for a randomly selected 14- to 18-year-old female is at most $210 be greater than or less than the probability for at most $200? Explain your thinking.

g.   Do you think that the probability that the monthly food cost for a randomly selected 14- to 18-year-old female is at most $210 is closer to 0.10, 0.30, or 0.50? Explain your thinking.

h.   Using a graphing calculator, and without using z-scores, find the probability (rounded to the nearest thousandth) that the monthly food cost for a randomly selected 14- to 18-year-old female is at most $210.

i.   Using a graphing calculator, and without using z-scores, find the probability (rounded to the nearest thousandth) that the monthly food cost for a randomly selected 14- to 18-year-old female is at least $235.

Example 3: Using a Spreadsheet to Find Normal Probabilities

Return to the information given in Example 1. The USDA, in its Official Food Plans (www.cnpp.usda.gov), states that the average cost of food for a 14- to 18-year-old male (on the Moderate-Cost Plan) is $261.50 per month. Assume that the monthly food cost for a 14- to 18-year-old male is approximately normally distributed with a mean of $261.50 and a standard deviation of $16.25. Round your answers to four decimal places.

Use a spreadsheet to find the probability that the monthly food cost for a randomly selected 14- to 18-year-old male is

a. Less than $280.

b. More than $270.

c. More than $250.

d. Between $240 and $275.

©2015 Great Minds. eureka-math.org
ALG II-M4-SE-B2-1.3.0-08.2015

Exercise 3

The USDA document described in Example 1 also states that the average cost of food for a 14- to 18-year-old female (again, on the Moderate-Cost Plan) is $215.20 per month. Assume that the monthly food cost for a 14- to 18-year-old female is approximately normally distributed with a mean of $215.20 and a standard deviation of $14.85. Round your answers to 4 decimal places.

Use a spreadsheet to find the probability that the monthly food cost for a randomly selected 14- to 18-year-old female is

    a.   Less than $225.

    b.   Less than $200.

    c.   More than $250.

    d.   Between $190 and $220.

Exercise 4

The reaction times of 490 people were measured. The results are shown in the frequency distribution below.

Reaction Time (seconds)	0.1 to < 0.15	0.15 to < 0.2	0.2 to < 0.25	0.25 to < 0.3	0.3 to < 0.35	0.35 to < 0.4
Frequency	9	82	220	138	37	4

a. Construct a histogram that displays these results.

b. Looking at the histogram, do you think a normal distribution would be an appropriate model for this distribution?

c. The mean of the reaction times for these 490 people is 0.2377, and the standard deviation of the reaction times is 0.0457. For a normal distribution with this mean and standard deviation, what is the probability that a randomly selected reaction time is at least 0.25?

d. The actual proportion of these 490 people who had a reaction time that was at least 0.25 is 0.365 (this can be calculated from the frequency distribution). How does this proportion compare to the probability that you calculated in part (c)? Does this confirm that the normal distribution is an appropriate model for the reaction time distribution?

Lesson Summary

Probabilities associated with normal distributions can be found using z-scores and tables of standard normal curve areas.

Probabilities associated with normal distributions can be found directly (without using z-scores) using a graphing calculator.

When a data distribution has a shape that is approximately normal, a normal distribution can be used as a model for the data distribution. The normal distribution with the same mean and the standard deviation as the data distribution is used.

Problem Set

1. Use a table of standard normal curve areas to find the following:

    a. The area to the left of $z = 1.88$

    b. The area to the right of $z = 1.42$

    c. The area to the left of $z = -0.39$

    d. The area to the right of $z = -0.46$

    e. The area between $z = -1.22$ and $z = -0.5$

2. Suppose that the durations of high school baseball games are approximately normally distributed with mean 105 minutes and standard deviation 11 minutes. Use a table of standard normal curve areas to find the probability that a randomly selected high school baseball game lasts

    a. Less than 115 minutes.

    b. More than 100 minutes.

    c. Between 90 and 110 minutes.

3. Using a graphing calculator, and *without* using z-scores, check your answers to Problem 2. (Round your answers to the nearest thousandth.)

4. In Problem 2, you were told that the durations of high school baseball games are approximately normally distributed with a mean of 105 minutes and a standard deviation of 11 minutes. Suppose also that the durations of high school softball games are approximately normally distributed with a mean of 95 minutes and the same standard deviation, 11 minutes. Is it more likely that a high school baseball game will last between 100 and 110 minutes or that a high school softball game will last between 100 and 110 minutes? Answer this question without doing any calculations.

5. A farmer has 625 female adult sheep. The sheep have recently been weighed, and the results are shown in the table below.

Weight (pounds)	140 to < 150	150 to < 160	160 to < 170	170 to < 180	180 to < 190	190 to < 200	200 to < 210
Frequency	8	36	173	221	149	33	5

a. Construct a histogram that displays these results.

b. Looking at the histogram, do you think a normal distribution would be an appropriate model for this distribution?

c. The weights of the 625 sheep have mean 174.21 pounds and standard deviation 10.11 pounds. For a normal distribution with this mean and standard deviation, what is the probability that a randomly selected sheep has a weight of at least 190 pounds? (Round your answer to the nearest thousandth.)

EUREKA
MATH

©2015 Great Minds. eureka-math.org
ALG II-M4-SE-B2-1.3.0-08.2015

Standard Normal Curve Areas

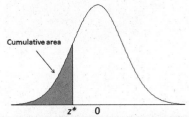

Cumulative area

z^* 0

z	0.00	0.01	0.02	0.03	0.04	0.05	0.06	0.07	0.08	0.09
−3.8	0.0001	0.0001	0.0001	0.0001	0.0001	0.0001	0.0001	0.0001	0.0001	0.0001
−3.7	0.0001	0.0001	0.0001	0.0001	0.0001	0.0001	0.0001	0.0001	0.0001	0.0001
−3.6	0.0002	0.0002	0.0001	0.0001	0.0001	0.0001	0.0001	0.0001	0.0001	0.0001
−3.5	0.0002	0.0002	0.0002	0.0002	0.0002	0.0002	0.0002	0.0002	0.0002	0.0002
−3.4	0.0003	0.0003	0.0003	0.0003	0.0003	0.0003	0.0003	0.0003	0.0003	0.0002
−3.3	0.0005	0.0005	0.0005	0.0004	0.0004	0.0004	0.0004	0.0004	0.0004	0.0003
−3.2	0.0007	0.0007	0.0006	0.0006	0.0006	0.0006	0.0006	0.0005	0.0005	0.0005
−3.1	0.0010	0.0009	0.0009	0.0009	0.0008	0.0008	0.0008	0.0008	0.0007	0.0007
−3.0	0.0013	0.0013	0.0013	0.0012	0.0012	0.0011	0.0011	0.0011	0.0010	0.0010
−2.9	0.0019	0.0018	0.0018	0.0017	0.0016	0.0016	0.0015	0.0015	0.0014	0.0014
−2.8	0.0026	0.0025	0.0024	0.0023	0.0023	0.0022	0.0021	0.0021	0.0020	0.0019
−2.7	0.0035	0.0034	0.0033	0.0032	0.0031	0.0030	0.0029	0.0028	0.0027	0.0026
−2.6	0.0047	0.0045	0.0044	0.0043	0.0041	0.0040	0.0039	0.0038	0.0037	0.0036
−2.5	0.0062	0.0060	0.0059	0.0057	0.0055	0.0054	0.0052	0.0051	0.0049	0.0048
−2.4	0.0082	0.0080	0.0078	0.0075	0.0073	0.0071	0.0069	0.0068	0.0066	0.0064
−2.3	0.0107	0.0104	0.0102	0.0099	0.0096	0.0094	0.0091	0.0089	0.0087	0.0084
−2.2	0.0139	0.0136	0.0132	0.0129	0.0125	0.0122	0.0119	0.0116	0.0113	0.0110
−2.1	0.0179	0.0174	0.0160	0.0166	0.0162	0.0158	0.0154	0.0150	0.0146	0.0143
−2.0	0.0228	0.0222	0.0217	0.0212	0.0207	0.0202	0.0197	0.0192	0.0188	0.0183
−1.9	0.0287	0.0281	0.0274	0.0268	0.0262	0.0256	0.0250	0.0244	0.0239	0.0233
−1.8	0.0359	0.0351	0.0344	0.0336	0.0329	0.0322	0.0314	0.0307	0.0301	0.0294
−1.7	0.0446	0.0436	0.0427	0.0418	0.0409	0.0401	0.0392	0.0384	0.0375	0.0367
−1.6	0.0548	0.0537	0.0526	0.0516	0.0505	0.0495	0.0485	0.0475	0.0465	0.0455
−1.5	0.0668	0.0655	0.0643	0.0630	0.0618	0.0606	0.0594	0.0582	0.0571	0.0599
−1.4	0.0808	0.0793	0.0778	0.0764	0.0749	0.0735	0.0721	0.0708	0.0694	0.0681
−1.3	0.0968	0.0951	0.0934	0.0918	0.0901	0.0885	0.0869	0.0853	0.0838	0.0823
−1.2	0.1151	0.1131	0.1112	0.1093	0.1075	0.1056	0.1038	0.1020	0.1003	0.0985
−1.1	0.1357	0.1335	0.1314	0.1292	0.1271	0.1251	0.1230	0.1210	0.1190	0.1170
−1.0	0.1587	0.1562	0.1539	0.1515	0.1492	0.1469	0.1446	0.1423	0.1401	0.1379
−0.9	0.1841	0.1814	0.1788	0.1762	0.1736	0.1711	0.1685	0.1660	0.1635	0.1611
−0.8	0.2119	0.2090	0.2061	0.2033	0.2005	0.1977	0.1949	0.1922	0.1894	0.1867
−0.7	0.2420	0.2389	0.2358	0.2327	0.2296	0.2266	0.2236	0.2206	0.2177	0.2148
−0.6	0.2743	0.2709	0.2676	0.2643	0.2611	0.2578	0.2546	0.2514	0.2483	0.2451
−0.5	0.3085	0.3050	0.3015	0.2981	0.2946	0.2912	0.2877	0.2843	0.2810	0.2776
−0.4	0.3446	0.3409	0.3372	0.3336	0.3300	0.3264	0.3228	0.3192	0.3156	0.3121
−0.3	0.3821	0.3783	0.3745	0.3707	0.3669	0.3632	0.3594	0.3557	0.3520	0.3483
−0.2	0.4207	0.4168	0.4129	0.4090	0.4052	0.4013	0.3974	0.3936	0.3897	0.3859
−0.1	0.4602	0.4562	0.4522	0.4483	0.4443	0.4404	0.4364	0.4325	0.4286	0.4247
−0.0	0.5000	0.4960	0.4920	0.4880	0.4840	0.4801	0.4761	0.4721	0.4681	0.4641

z	0.00	0.01	0.02	0.03	0.04	0.05	0.06	0.07	0.08	0.09
0.0	0.5000	0.5040	0.5080	0.5120	0.5160	0.5199	0.5239	0.5279	0.5319	0.5359
0.1	0.5398	0.5438	0.5478	0.5517	0.5557	0.5596	0.5636	0.5675	0.5714	0.5753
0.2	0.5793	0.5832	0.5871	0.5910	0.5948	0.5987	0.6026	0.6064	0.6103	0.6141
0.3	0.6179	0.6217	0.6255	0.6293	0.6331	0.6368	0.6406	0.6443	0.6480	0.6517
0.4	0.6554	0.6591	0.6628	0.6664	0.6700	0.6736	0.6772	0.6808	0.6844	0.6879
0.5	0.6915	0.6950	0.6985	0.7019	0.7054	0.7088	0.7123	0.7157	0.7190	0.7224
0.6	0.7257	0.7291	0.7324	0.7357	0.7389	0.7422	0.7454	0.7486	0.7517	0.7549
0.7	0.7580	0.7611	0.7642	0.7673	0.7704	0.7734	0.7764	0.7794	0.7823	0.7852
0.8	0.7881	0.7910	0.7939	0.7967	0.7995	0.8023	0.8051	0.8078	0.8106	0.8133
0.9	0.8159	0.8186	0.8212	0.8238	0.8264	0.8289	0.8315	0.8340	0.8365	0.8389
1.0	0.8413	0.8438	0.8461	0.8485	0.8508	0.8531	0.8554	0.8577	0.8599	0.8621
1.1	0.8643	0.8665	0.8686	0.8708	0.8729	0.8749	0.8770	0.8790	0.8810	0.8830
1.2	0.8849	0.8869	0.8888	0.8907	0.8925	0.8944	0.8962	0.8980	0.8997	0.9015
1.3	0.9032	0.9049	0.9066	0.9082	0.9099	0.9115	0.9131	0.9147	0.9162	0.9177
1.4	0.9192	0.9207	0.9222	0.9236	0.9251	0.9265	0.9279	0.9292	0.9306	0.9319
1.5	0.9332	0.9345	0.9357	0.9370	0.9382	0.9394	0.9406	0.9418	0.9429	0.9441
1.6	0.9452	0.9463	0.9474	0.9484	0.9495	0.9505	0.9515	0.9525	0.9535	0.9545
1.7	0.9554	0.9564	0.9573	0.9582	0.9591	0.9599	0.9608	0.9616	0.9625	0.9633
1.8	0.9641	0.9649	0.9656	0.9664	0.9671	0.9678	0.9686	0.9693	0.9699	0.9706
1.9	0.9713	0.9719	0.9726	0.9732	0.9738	0.9744	0.9750	0.9756	0.9761	0.9767
2.0	0.9772	0.9778	0.9783	0.9788	0.9793	0.9798	0.9803	0.9808	0.9812	0.9817
2.1	0.9821	0.9826	0.9830	0.9834	0.9838	0.9842	0.9846	0.9850	0.9854	0.9857
2.2	0.9861	0.9864	0.9868	0.9871	0.9875	0.9878	0.9881	0.9884	0.9887	0.9890
2.3	0.9893	0.9896	0.9898	0.9901	0.9904	0.9906	0.9909	0.9911	0.9913	0.9916
2.4	0.9918	0.9920	0.9922	0.9925	0.9927	0.9929	0.9931	0.9932	0.9934	0.9936
2.5	0.9938	0.9940	0.9941	0.9943	0.9945	0.9946	0.9948	0.9949	0.9951	0.9952
2.6	0.9953	0.9955	0.9956	0.9957	0.9959	0.9960	0.9961	0.9962	0.9963	0.9964
2.7	0.9965	0.9966	0.9967	0.9968	0.9969	0.9970	0.9971	0.9972	0.9973	0.9974
2.8	0.9974	0.9975	0.9976	0.9977	0.9977	0.9978	0.9979	0.9979	0.9980	0.9981
2.9	0.9981	0.9982	0.9982	0.9983	0.9984	0.9984	0.9985	0.9985	0.9986	0.9986
3.0	0.9987	0.9987	0.9987	0.9988	0.9988	0.9989	0.9989	0.9989	0.9990	0.9990
3.1	0.9990	0.9991	0.9991	0.9991	0.9992	0.9992	0.9992	0.9992	0.9993	0.9993
3.2	0.9993	0.9993	0.9994	0.9994	0.9994	0.9994	0.9994	0.9995	0.9995	0.9995
3.3	0.9995	0.9995	0.9995	0.9996	0.9996	0.9996	0.9996	0.9996	0.9996	0.9997
3.4	0.9997	0.9997	0.9997	0.9997	0.9997	0.9997	0.9997	0.9997	0.9997	0.9998
3.5	0.9998	0.9998	0.9998	0.9998	0.9998	0.9998	0.9998	0.9998	0.9998	0.9998
3.6	0.9998	0.9998	0.9999	0.9999	0.9999	0.9999	0.9999	0.9999	0.9999	0.9999
3.7	0.9999	0.9999	0.9999	0.9999	0.9999	0.9999	0.9999	0.9999	0.9999	0.9999
3.8	0.9999	0.9999	0.9999	0.9999	0.9999	0.9999	0.9999	0.9999	0.9999	0.9999

Lesson 12: Types of Statistical Studies

Classwork

Opening Exercise

You want to know what proportion of the population likes rock music. You carefully consider three ways to conduct a study. What are the similarities and differences between the following three alternatives? Do any display clear advantages or disadvantages over the others?

a. You could pick a random sample of people and ask them the question, "Do you like rock music?" and record their answers.

b. You could pick a random sample of people and follow them for a period of time, noting their music purchases, both in stores and online.

c. You could pick a random sample of people, separate them into groups, and have each group listen to a different genre of music. You would collect data on the people who display an emotional response to the rock music.

A statistical study begins by asking a question that can be answered with data. The next steps are to collect appropriate data, organize and analyze them, and arrive at a conclusion in the context of the original question. This lesson focuses on the three main types of statistical studies: observational studies, surveys, and experiments. The objective of an observational study and a survey is to learn about characteristics of some population, so the data should be collected in a way that would result in a representative sample. This speaks to the importance of random selection of subjects for the study. The objective of an experiment is to answer such questions as "What is the effect of treatments on a response variable?" Data in an experiment need to be collected in a way that does not favor one treatment over another. This demonstrates the importance of random assignment of subjects in the study to the treatments.

An observational study is one in which the values of one or more variables are observed with no attempt to affect the outcomes. One kind of observational study is a survey. A survey requires asking a group of people to respond to one or more questions. (A poll is one example of a survey.) An experiment differs from an observational study. In an experiment, subjects are assigned to treatments for the purpose of seeing what effect the treatment has on some response while an observational study makes no attempt to affect the outcomes (i.e., no treatment is given). Note that subjects could be people, animals, or any set of items that produce variability in their responses. Here is an example of an observational study: In a random sample of students, it was observed that those students who played a musical instrument had better grades than those who did not play a musical instrument. In an experiment, a group of students who do not currently play a musical instrument would be assigned at random to having to play a musical instrument or not having to play a musical instrument for a certain period of time. Then, at the end of the period of time, we would compare academic performance.

Classify each of the three study methods about rock music as an observational study, a survey, or an experiment.

©2015 Great Minds. eureka-math.org
ALG II-M4-SE-B2-1.3.0-08.2015

Example 1: Survey

Item	I like the item.	I do not like the item.	I have never tried the item.
Salad			
Vegetable Pizza			
Turkey Sandwich			
Raspberry Tea			

a. It is easy to determine if a study is a survey. A survey asks people to respond to questions. But surveys can be flawed in several ways. Questions may be confusing. For example, consider the following question:

> What kind of computer do you own? (*Circle one*)   Mac   IBM-PC

How do you answer that question if you do not own a computer? How do you answer that question if you own a different brand? A better question would be

> Do you own a computer? (*Circle one*) Yes   No

> If you answered yes, what brand of computer is it? _____

Now consider the question, "Do you like your school's cafeteria food?"

Rewrite the question in a better form. Keep in mind that not all students may use the school's cafeteria, and even if they do, there may be some foods that they like and some that they do not like.

b. Something else to consider with surveys is how survey participants are chosen. If the purpose of the survey is to learn about some population, ideally participants would be randomly selected from the population of interest. If people are not randomly selected, misleading conclusions from the survey data may be drawn. There are many famous examples of this. Perhaps the most famous case was in 1936 when *The Literary Digest* magazine predicted that Alf Landon would beat incumbent President Franklin Delano Roosevelt by 370 electoral votes to 161. Roosevelt won 523 to 8.

Ten million questionnaires were sent to prospective voters (selected from the magazine's subscription list, automobile registration lists, phone lists, and club membership lists), and over two million questionnaires were returned. Surely such a large sample should represent the whole population. How could *The Literary Digest* prediction be so far off the mark?

c. Write or say to your neighbor two things that are important about surveys.

Example 2: Observational Study

a. An observational study records the values of variables for members of a sample but does not attempt to influence the responses. For example, researchers investigated the link between the use of cell phones and brain cancer. There are two variables in this study: One is the extent of cell phone usage, and the second is whether a person has brain cancer. Both variables were measured for a group of people. This is an observational study. There was no attempt to influence peoples' cell phone usage to see if different levels of usage made any difference in whether or not a person developed brain cancer.

Why would studying any relationship between asbestos exposure and lung cancer be an observational study and not an experiment?

b. In an observational study (just as in surveys), the people or objects to be observed would ideally be selected at random from the population of interest. This would eliminate bias and make it possible to generalize from a sample to a population. For example, to determine if the potato chips made in a factory contain the desired amount of salt, a sample of chips would be selected randomly so that the sample can be considered to be representative of the population of chips.

Discuss how a random sample of 100 chips might be selected from a conveyor belt of chips.

c. Suppose that an observational study establishes a link between asbestos exposure and lung cancer. Based on that finding, can we conclude that asbestos exposure causes lung cancer? Why or why not?

d. Write or say to your neighbor two things that are important about observational studies.

Example 3: Experiment

a. An experiment imposes treatments to see the effect of the treatments on some response. Suppose that an observational study indicated that a certain type of tree did not have as much termite damage as other trees. Researchers wondered if resin from the tree was toxic to termites. They decided to do an experiment where they exposed some termites to the resin and others to plain water and recorded whether the termites survived. The explanatory variable (treatment variable) is the exposure type (resin, plain water), and the response variable is whether or not the termites survived. We know this is an experiment because the researchers imposed a treatment (exposure type) on the subjects (termites).

Is the following an observational study or an experiment? Why? If it is an experiment, identify the treatment variable and the response variable. If it is an observational study, identify the population of interest.

A study was done to answer the question, "What is the effect of different durations of light and dark on the growth of radish seedlings?" Three similar growth chambers (plastic bags) were created in which 30 seeds randomly chosen from a package were placed in each chamber. One chamber was randomly selected and placed in 24 hours of light, another for 12 hours of light and 12 hours of darkness, and a third for 24 hours of darkness. After three days, researchers measured and recorded the lengths of radish seedlings for the germinating seeds.

b. In an experiment, random assignment of subjects to treatments is done to create comparable treatment groups. For example, a university biologist wants to compare the effects of two weed killers on pansies. She chooses 24 plants. If she applies weed killer A to the 12 healthiest plants and B to the remaining 12 plants, she will not know which plants died due to the type of weed killer used and which plants subjected to weed killer B were already on their last legs. Randomly selecting 12 plants to receive weed killer A and then assigning the rest to B would help ensure that the plants in each group are fairly similar.

How might the biologist go about randomly assigning 12 plants from the 24 candidates to receive weed killer A? Could she be sure to get exactly 12 plants assigned to weed killer A and 12 plants to weed killer B by tossing a fair coin for each plant and assigning "heads up" plants to weed killer A and "tails up" to weed killer B? If not, suggest a method that you would use.

c. Write or say to your neighbor two things that are important about experiments.

Lesson 12:   Types of Statistical Studies

Exercises 1–3

1. For each of the following study descriptions, identify whether the study is a survey, an observational study, or an experiment, and give a reason for your answer. For observational studies, identify the population of interest. For experiments, identify the treatment and response variables.

   a. A study investigated whether boys are quicker at learning video games than girls. Twenty randomly selected boys and twenty randomly selected girls played a video game that they had never played before. The time it took them to reach a certain level of expertise was recorded.

   b. As your statistics project, you collect data by posting five questions on poster board around your classroom and recording how your classmates respond to them.

   c. A professional sports team traded its best player. The local television station wanted to find out what the fans thought of the trade. At the beginning of the evening news program, they asked viewers to call one number if they favored the trade and a different number if they were opposed to the trade. At the end of the news program, they announced that 53.7% of callers favored the trade.

   d. The local department of transportation is responsible for maintaining lane and edge lines on its paved roads. There are two new paint products on the market. Twenty comparable stretches of road are identified. Paint A is randomly assigned to ten of the stretches of road and paint B to the other ten. The department finds that paint B lasts longer.

   e. The National Highway Traffic Safety Administration conducts annual studies on drivers' seatbelt use at a random selection of roadway sites in each state in the United States. To determine if seatbelt usage has increased, data are analyzed over two successive years.

f. People should brush their teeth at least twice a day for at least two to three minutes with each brushing. For a statistics class project, you ask a random number of students at your school questions concerning their tooth brushing activities.

g. A study determines whether taking aspirin regularly helps to prevent heart attacks. A large group of male physicians of comparable health were randomly assigned equally to taking an aspirin every second day or to taking a placebo. After several years, the proportion of the study participants who had suffered heart attacks in each group was compared.

2. For the following, is the stated conclusion reasonable? Why or why not?

A study found a positive relationship between the happiness of elderly people and the number of pets they have. Therefore, having more pets causes elderly people to be happier.

3. A researcher wanted to find out whether higher levels of a certain drug given to experimental rats would decrease the time it took them to complete a given maze to find food.

a. Why would the researcher have to carry out an experiment rather than an observational study?

b. Describe an experiment that the researcher might carry out based on 30 comparable rats and three dosage levels: 0 mg, 1 mg, and 2 mg.

Lesson Summary

- There are three major types of statistical studies: observational studies, surveys, and experiments.
    - An *observational study* records the values of variables for members of a sample.
    - A *survey* is a type of observational study that gathers data by asking people a number of questions.
    - An *experiment* assigns subjects to treatments for the purpose of seeing what effect the treatments have on some response.
- To avoid bias in observational studies and surveys, it is important to select subjects randomly.
- Cause-and-effect conclusions cannot be made in observational studies or surveys.
- In an experiment, it is important to assign subjects to treatments randomly in order to make cause-and-effect conclusions.

Problem Set

1. State if the following is an observational study, a survey, or an experiment, and give a reason for your answer.

   Linda wanted to know if it is easier for students to memorize a list of common three-letter words (such as *fly, pen,* and *red*) than a list of three-letter nonsense words (such as *vir, zop,* and *twq*). She randomly selected 28 students from all tenth graders in her district. She put 14 blue and 14 red chips in a jar, and without looking, each student chose a chip. Those with red chips were given the list of common words; those with blue chips were given the list of nonsense words. She gave all students one minute to memorize their lists. After the minute, she collected the lists and asked the students to write down all the words that they could remember. She recorded the number of correct words recalled.

2. State if the following is an observational study, a survey, or an experiment, and give a reason for your answer.

   Ken wants to compare how many hours a week that sixth graders spend doing mathematics homework to how many hours a week that eleventh graders spend doing mathematics homework. He randomly selects ten sixth graders and ten eleventh graders and records how many hours each student spent on mathematics homework in a certain week.

3. Suppose that in your health class you read two studies on the relationship between eating breakfast and success in school for elementary school children. Both studies concluded that eating breakfast causes elementary school children to be successful in school.

   a. Suppose that one of the studies was an observational study. Describe how you would recognize that they had conducted an observational study. Were the researchers correct in their causal conclusion?

   b. Suppose that one of the studies was an experiment. Describe how you would recognize that they had conducted an experiment. Were the researchers correct in their causal conclusion?

4. Data from a random sample of 50 students in a school district showed a positive relationship between reading score on a standardized reading exam and shoe size. Can it be concluded that having bigger feet causes one to have a higher reading score? Explain your answer.

Use the following scenarios for Problems 5–7.

A. Researchers want to determine if there is a relationship between whether or not a woman smoked during pregnancy and the birth weight of her baby. Researchers examined records for the past five years at a large hospital.

B. A large high school wants to know the proportion of students who currently use illegal drugs. Uniformed police officers asked a random sample of 200 students about their drug use.

C. A company develops a new dog food. The company wants to know if dogs would prefer its new food over the competition's dog food. One hundred dogs, who were food deprived overnight, were given equal amounts of the two dog foods: the new food versus the competitor's food. The proportion of dogs preferring the new food versus the competitor's was recorded.

5. Which scenario above describes an experiment? Explain why.

6. Which scenario describes a survey? Will the results of the survey be accurate? Why or why not?

7. The remaining scenario is an observational study. Is it possible to perform an experiment to determine if a relationship exists? Why or why not?

Lesson 13: Using Sample Data to Estimate a Population Characteristic

Classwork

Example 1: Population and Sample

Answer the following questions, and then share your responses with a neighbor.

a. A team of scientists wants to determine the average length and weight of fish in Lake Lucerne. Name a sample that can be used to help answer their question.

b. Golf balls from different manufacturers are tested to determine which brand travels the farthest. What is the population being studied?

Exercise 1

For each of the following, does the group described constitute a population or a sample? Or could it be considered to be either a population or a sample? Explain your answer.

a. The animals that live in Yellowstone National Park

b. The first-run movies released last week that were shown at the local theater complex last weekend

c. People who are asked how they voted in an exit poll

d. Some cars on the lot of the local car dealer

e. The words of the Gettysburg Address

f. The colors of pencils available in a 36-count packet of colored pencils

g. The students from your school who attended your school's soccer game yesterday

Example 2: Representative Sample

If a sample is taken for the purpose of generalizing to a population, the sample must be representative of the population. In other words, it must be similar to the population even though it is smaller than the population. For example, suppose you are the campaign manager for your friend who is running for senior class president. You would like to know what proportion of students would vote for her if the election was held today. The class is too big to ask everyone (314 students). What would you do?

Comment on whether or not each of the following sampling procedures should be used. Explain why or why not.

a. Poll everyone in your friend's math class.

b. Assign every student in the senior class a number from 1 to 314. Then, use a random number generator to select 30 students to poll.

c. Ask every student who is going through the lunch line in the cafeteria who they will vote for.

Exercise 2

There is no procedure that guarantees a representative sample. But the best procedure to obtain a representative sample is one that gives every different possible sample an equal chance to be chosen. The sample resulting from such a procedure is called a *random sample*.

Suppose that you want to randomly select 60 employees from a group of 625 employees.

Explain how to use a random number table or a calculator with a random number generator to choose 60 different numbers at random and include the students with these numbers in the sample.

Example 3: Population Characteristics and Sample Statistics

A statistical study begins with a question of interest that can be answered by data. Depending on the study, data could be collected from all individuals in the population or from a random sample of individuals selected from the population. Read through the following, and identify which of the summary measures represents a *population characteristic* and which represents a *sample statistic*. Explain your reasoning for each.

Suppose the population of interest is the words of the Gettysburg Address. There are 269 of them (depending on the version).

a. The proportion of nouns in all words of the Gettysburg Address

b. The proportion of nouns or the proportion of words containing the letter *e* in a random sample of words taken from the Gettysburg Address

c. The mean length of the words in a random sample of words taken from the Gettysburg Address

d. The proportion of all words in the Gettysburg Address that contain the letter *e*

e. The mean length of all words in the Gettysburg Address

Exercise 3

For the following items of interest, describe an appropriate population, population characteristic, sample, and sample statistic. Explain your answer.

    a.   Time it takes students to run a quarter mile

    b.   National forests that contain bald eagle nests

    c.   Curfew time of boys compared to girls

    d.   Efficiency of electric cars

©2015 Great Minds. eureka-math.org
ALG II-M4-SE-B2-1.3.0-08.2015

Exercise 4

Consider the following questions:

- What proportion of eleventh graders at our high school are taking at least one advanced placement course?
- What proportion of eleventh graders at our high school have a part-time job?
- What is the typical number of hours an eleventh grader at our high school studies outside of school hours on a weekday (Monday, Tuesday, Wednesday, or Thursday)?
- What is the typical time (in minutes) that students at our high school spend getting to school?
- What is the proportion of students at our high school who plan to attend a college or technical school after graduation?
- What is the typical amount of time (in hours per week) that students at our high school are involved in community service?

Select one of these questions (or a different statistical question that has been approved by your teacher). Working with your group, write a paragraph that:

- States the statistical question of interest pertaining to the students in the population for the statistical question selected.
- Identifies a population characteristic of interest.
- Identifies the appropriate statistic based on a sample of 40 students.
- States what property your sample must have for you to be able to use its results to generalize to all students in your high school.
- Includes the details on how you would select your sample.

Lesson Summary

We refer to summary measures calculated using data from an entire population as *population characteristics*. We refer to summary measures calculated using data from a sample as *sample statistics*. To generalize from a sample to the corresponding population, it is important that the sample be a random sample from the population. A *random sample* is one that is selected in a way that gives every different possible sample an equal chance of being chosen.

Problem Set

1. In the following, identify whether the subjects being measured are the sample or the population. In some cases, they could be considered a sample or a population. Explain each answer.

Subjects	What Is Being Measured	Sample or Population? Explain
Some students in your class	Number of books in backpack	
AA batteries of a certain brand	Lifetime	
Birds in Glacier National Park	Number of species	
Students in your school	Number absent or present today	
Words in the Constitution of the U.S.	Whether a noun or not	
Americans of voting age	Opinion on an issue	

2. For the following items of interest, describe an appropriate population, a population characteristic, a sample, and a sample statistic:

   a. Whether or not a driver is speeding in your school zone during school hours in a day

   b. Seatbelt usage of men compared to women

   c. Impact of a new antidepressant on people with severe headaches

3. What are the identification numbers for ten students chosen at random from a population of 78 students based on the following string of random digits? Start at the left.

   27816 78416 01822 73521 37741 016312 68000 53645 56644 97892 63408 77919 44575

This page intentionally left blank

Table of Random Digits

Row																				
1	6	6	7	2	8	0	0	8	4	0	0	4	6	0	3	2	2	4	6	8
2	8	0	3	1	1	1	1	2	7	0	1	9	1	2	7	1	3	3	5	3
3	5	3	5	7	3	6	3	1	7	2	5	5	1	4	7	1	6	5	6	5
4	9	1	1	9	2	8	3	0	3	6	7	7	4	7	5	9	8	1	8	3
5	9	0	2	9	9	7	4	6	3	6	6	3	7	4	2	7	0	0	1	9
6	8	1	4	6	4	6	8	2	8	9	5	5	2	9	6	2	5	3	0	3
7	4	1	1	9	7	0	7	2	9	0	9	7	0	4	6	2	3	1	0	9
8	9	9	2	7	1	3	2	9	0	3	9	0	7	5	6	7	1	7	8	7
9	3	4	2	2	9	1	9	0	7	8	1	6	2	5	3	9	0	9	1	0
10	2	7	3	9	5	9	9	3	2	9	3	9	1	9	0	5	5	1	4	2
11	0	2	5	4	0	8	1	7	0	7	1	3	0	4	3	0	6	4	4	4
12	8	6	0	5	4	8	8	2	7	7	0	1	0	1	7	1	3	5	3	4
13	4	2	6	4	5	2	4	2	6	1	7	5	6	6	4	0	8	4	1	2
14	4	4	9	8	7	3	4	3	8	2	9	1	5	3	5	9	8	9	2	9
15	6	4	8	0	0	0	4	2	3	8	1	8	4	0	9	5	0	9	0	4
16	3	2	3	8	4	8	8	6	2	9	1	0	1	9	9	3	0	7	3	5
17	6	6	7	2	8	0	0	8	4	0	0	4	6	0	3	2	2	4	6	8
18	8	0	3	1	1	1	1	2	7	0	1	9	1	2	7	1	3	3	5	3
19	5	3	5	7	3	6	3	1	7	2	5	5	1	4	7	1	6	5	6	5
20	9	1	1	9	2	8	3	0	3	6	7	7	4	7	5	9	8	1	8	3
21	9	0	2	9	9	7	4	6	3	6	6	3	7	4	2	7	0	0	1	9
22	8	1	4	6	4	6	8	2	8	9	5	5	2	9	6	2	5	3	0	3
23	4	1	1	9	7	0	7	2	9	0	9	7	0	4	6	2	3	1	0	9
24	9	9	2	7	1	3	2	9	0	3	9	0	7	5	6	7	1	7	8	7
25	3	4	2	2	9	1	9	0	7	8	1	6	2	5	3	9	0	9	1	0
26	2	7	3	9	5	9	9	3	2	9	3	9	1	9	0	5	5	1	4	2
27	0	2	5	4	0	8	1	7	0	7	1	3	0	4	3	0	6	4	4	4
28	8	6	0	5	4	8	8	2	7	7	0	1	0	1	7	1	3	5	3	4
29	4	2	6	4	5	2	4	2	6	1	7	5	6	6	4	0	8	4	1	2
30	4	4	9	8	7	3	4	3	8	2	9	1	5	3	5	9	8	9	2	9
31	6	4	8	0	0	0	4	2	3	8	1	8	4	0	9	5	0	9	0	4
32	3	2	3	8	4	8	8	6	2	9	1	0	1	9	9	3	0	7	3	5
33	6	6	7	2	8	0	0	8	4	0	0	4	6	0	3	2	2	4	6	8
34	8	0	3	1	1	1	1	2	7	0	1	9	1	2	7	1	3	3	5	3
35	5	3	5	7	3	6	3	1	7	2	5	5	1	4	7	1	6	5	6	5
36	9	1	1	9	2	8	3	0	3	6	7	7	4	7	5	9	8	1	8	3
37	9	0	2	9	9	7	4	6	3	6	6	3	7	4	2	7	0	0	1	9
38	8	1	4	6	4	6	8	2	8	9	5	5	2	9	6	2	5	3	0	3
39	4	1	1	9	7	0	7	2	9	0	9	7	0	4	6	2	3	1	0	9
40	9	9	2	7	1	3	2	9	0	3	9	0	7	5	6	7	1	7	8	7

This page intentionally left blank

Lesson 14: Sampling Variability in the Sample Proportion

Classwork

Example 1: Polls

A recent poll stated that 40% of Americans pay "a great deal" or a "fair amount" of attention to the nutritional information that restaurants provide. This poll was based on a random sample of 2,027 adults living in the United States.

The 40% corresponds to a proportion of 0.40, and 0.40 is called a *sample proportion*. It is an estimate of the proportion of all adults who would say they pay "a great deal" or a "fair amount" of attention to the nutritional information that restaurants provide. If you were to take a random sample of 20 Americans, how many would you predict would say that they pay attention to nutritional information? In this lesson, you will investigate this question by generating distributions of sample proportions and investigating patterns in these distributions.

Your teacher will give your group a container of dried beans. Some of the beans in the container are black. With your classmates, you are going to see what happens when you take a sample of beans from the container and use the proportion of black beans in the sample to estimate the proportion of black beans in the container (a *population proportion*).

Exploratory Challenge 1/Exercises 1–9

1. Each person in the group should randomly select a sample of 20 beans from the container by carefully mixing all the beans and then selecting one bean and recording its color. Replace the bean, mix the bag, and continue to select one bean at a time until 20 beans have been selected. Be sure to replace each bean and mix the bag before selecting the next bean. Count the number of black beans in your sample of 20.

2. What is the proportion of black beans in your sample of 20? (Round your answer to 2 decimal places.) This value is called the *sample proportion* of black beans.

3. Write your sample proportion on a sticky note, and place the note on the number line that your teacher has drawn on the board. Place your note above the value on the number line that corresponds to your sample proportion.

The graph of all the students' sample proportions is called the *sampling distribution* of the samples' proportions. This sampling distribution is an approximation of the actual sampling distribution of all possible samples of size 20.

4. Describe the shape of the distribution.

5. What was the smallest sample proportion observed?

6. What was the largest sample proportion observed?

7. What sample proportion occurred most often?

8. Using technology, find the mean and standard deviation of the sample proportions used to construct the sampling distribution created by the class.

9. How does the mean of the sampling distribution compare with the population proportion of 0.40?

Example 2: Sampling Variability

What do you think would happen to the sampling distribution if everyone in class took a random sample of 40 beans from the container? To help answer this question, you will repeat the process described in Example 1, but this time you will draw a random sample of 40 beans instead of 20.

Exploratory Challenge 2/Exercises 10–21

10. Take a random sample with replacement of 40 beans from the container. Count the number of black beans in your sample of 40 beans.

11. What is the proportion of black beans in your sample of 40? (Round your answer to 2 decimal places.)

12. Write your sample proportion on a sticky note, and place it on the number line that your teacher has drawn on the board. Place your note above the value on the number line that corresponds to your sample proportion.

13. Describe the shape of the distribution.

14. What was the smallest sample proportion observed?

15. What was the largest sample proportion observed?

16. What sample proportion occurred most often?

17. Using technology, find the mean and standard deviation of the sample proportions used to construct the sampling distribution created by the class.

18. How does the mean of the sampling distribution compare with the population proportion of 0.40?

19. How does the mean of the sampling distribution based on random samples of size 20 compare to the mean of the sampling distribution based on random samples of size 40?

20. As the sample size increased from 20 to 40, describe what happened to the sampling variability (standard deviation of the distribution of sample proportions)?

21. What do you think would happen to the variability (standard deviation) of the distribution of sample proportions if the sample size for each sample was 80 instead of 40? Explain.

EUREKA
MATH™

Lesson Summary

The sampling distribution of the sample proportion can be approximated by a graph of the sample proportions for many different random samples. The mean of the sampling distribution of the sample proportions will be approximately equal to the value of the population proportion.

As the sample size increases, the sampling variability in the sample proportion decreases; in other words, the standard deviation of the sampling distribution of the sample proportions decreases.

Problem Set

1. A class of 28 eleventh graders wanted to estimate the proportion of all juniors and seniors at their high school with part-time jobs after school. Each eleventh grader took a random sample of 30 juniors and seniors and then calculated the proportion with part-time jobs. Following are the 28 sample proportions.

    0.7, 0.8, 0.57, 0.63, 0.7, 0.47, 0.67, 0.67, 0.8, 0.77, 0.4, 0.73, 0.63, 0.67, 0.6, 0.77, 0.77, 0.77, 0.53, 0.57, 0.73, 0.7, 0.67, 0.7, 0.77, 0.57, 0.77, 0.67

    a. Construct a dot plot of the sample proportions.

    b. Describe the shape of the distribution.

    c. Using technology, find the mean and standard deviation of the sample proportions.

    d. Do you think that the proportion of all juniors and seniors at the school with part-time jobs could be 0.7? Do you think it could be 0.5? Justify your answers based on your dot plot.

    e. Suppose the eleventh graders had taken random samples of size 60. How would the distribution of sample proportions based on samples of size 60 differ from the distribution for samples of size 30?

2. A group of eleventh graders wanted to estimate the proportion of all students at their high school who suffer from allergies. Each student in one group of eleventh graders took a random sample of 20 students, while each student in another group of eleventh graders each took a random sample of 40 students. Below are the two sampling distributions (shown as histograms) of the sample proportions of high school students who said that they suffer from allergies. Which histogram is based on random samples of size 40? Explain.

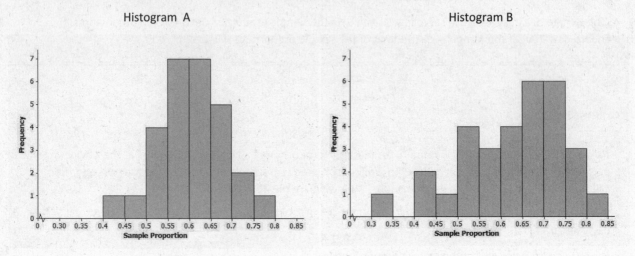

3. The nurse in your school district would like to study the proportion of all high school students in the district who usually get at least eight hours of sleep on school nights. Suppose each student in your class takes a random sample of 20 high school students in the district and each calculates their sample proportion of students who said that they usually get at least eight hours of sleep on school nights. Below is a histogram of the sampling distribution.

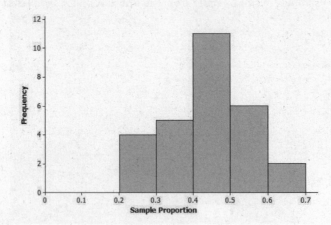

a. Do you think that the proportion of all high school students who usually get at least eight hours of sleep on school nights could have been 0.4? Do you think it could have been 0.55? Could it have been 0.75? Justify your answers based on the histogram.

b. Suppose students had taken random samples of size 60. How would the distribution of sample proportions based on samples of size 60 differ from those of size 20?

Lesson 15: Sampling Variability in the Sample Proportion

Classwork

Example 1

A high school principal claims that 50% of the school's students walk to school in the morning. A student attempts to verify the principal's claim by taking a random sample of 40 students and asking them if they walk to school in the morning. Sixteen of the sampled students say they usually walk to school in the morning, giving a sample proportion of $\frac{16}{40} = 0.40$, which seems to dispel the principal's claim of 50%. But could the principal be correct that the proportion of all students who walk to school is 50%?

a. Make a conjecture about the answer.

b. Develop a plan for how to respond.

Help the student make a decision on the principal's claim by investigating what kind of sample proportions you would expect to see if the principal's claim of 50% is true. You will do this by using technology to simulate the flipping of a coin 40 times.

Exploratory Challenge 1/Exercises 1–9

In Exercises 1–9, students should assume that the principal is correct that 50% of the population of students walk to school. Designate heads to represent a student who walks to school.

1. Simulate 40 flips of a fair coin. Record your observations in the space below.

2. What is the sample proportion of heads in your sample of 40? Report this value to your teacher.

3. Repeat Exercises 1 and 2 to obtain a second sample of 40 coin flips.

Your teacher will display a graph of all the students' sample proportions of heads.

4. Describe the shape of the distribution.

5. What was the smallest sample proportion observed?

6. What was the largest sample proportion observed?

EUREKA
MATH™

7. Estimate the center of the distribution of sample proportions.

Your teacher will report the mean and standard deviation of the sampling distribution created by the class.

8. How does the mean of the sampling distribution compare with the population proportion of 0.50?

9. Recall that a student took a random sample of 40 students and found that the sample proportion of students who walk to school was 0.40. Would this have been a surprising result if the actual population proportion was 0.50 as the principal claims?

Example 2: Sampling Variability

What do you think would happen to the sampling distribution you constructed in the previous exercises had everyone in class taken a random sample of size 80 instead of 40? Justify your answer. This will be investigated in the following exercises.

Exploratory Challenge 2/Exercises 10–22

10. Use technology and simulate 80 coin flips. Calculate the proportion of heads. Record your results in the space below.

11. Repeat flipping a coin 80 times until you have recorded a total of 40 sample proportions.

12. Construct a dot plot of the 40 sample proportions.

13. Describe the shape of the distribution.

14. What was the smallest proportion of heads observed?

15. What was the largest proportion of heads observed?

16. Using technology, find the mean and standard deviation of the distribution of sample proportions.

17. Compare your results with the others in your group. Did you have similar means and standard deviations?

©2015 Great Minds. eureka-math.org
ALG II-M4-SE-B2-1.3.0-08.2015

18. How does the mean of the sampling distribution based on 40 simulated flips of a coin (Exercise 1) compare to the mean of the sampling distribution based on 80 simulated coin flips?

19. Describe what happened to the sampling variability (standard deviation) of the distribution of sample proportions as the number of simulated coin flips increased from 40 to 80.

20. What do you think would happen to the variability (standard deviation) of the distribution of sample proportions if the sample size for each sample was 200 instead of 80? Explain.

21. Recall that a student took a random sample of 40 students and found that the sample proportion of students who walk to school was 0.40. If the student had taken a random sample of 80 students instead of 40, would this have been a surprising result if the actual population proportion was 0.50 as the principal claims?

22. What do you think would happen to the sampling distribution you constructed in the previous exercises if everyone in class took a random sample of size 80 instead of 40? Justify your answer.

Lesson Summary

The sampling distribution of the sample proportion can be approximated by a graph of the sample proportions for many different random samples. The mean of the sample proportions will be approximately equal to the value of the population proportion.

As the sample size increases, the sampling variability in the sample proportion decreases; in other words, the standard deviation of the sample proportions decreases.

Problem Set

1. A student conducted a simulation of 30 coin flips. Below is a dot plot of the sampling distribution of the proportion of heads. This sampling distribution has a mean of 0.51 and a standard deviation of 0.09.

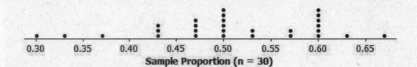

   a. Describe the shape of the distribution.

   b. Describe what would have happened to the mean and the standard deviation of the sampling distribution of the sample proportions if the student had flipped a coin 50 times, calculated the proportion of heads, and then repeated this process for a total of 30 times.

2. What effect does increasing the sample size have on the mean of the sampling distribution?

3. What effect does increasing the sample size have on the standard deviation of the sampling distribution?

4. A student wanted to decide whether or not a particular coin was fair (i.e., the probability of flipping a head is 0.5). She flipped the coin 20 times, calculated the proportion of heads, and repeated this process a total of 40 times. Below is the sampling distribution of sample proportions of heads. The mean and standard deviation of the sampling distribution are 0.379 and 0.091, respectively. Do you think this was a fair coin? Why or why not?

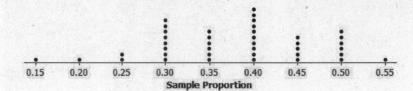

5. The same student flipped the coin 100 times, calculated the proportion of heads, and repeated this process a total of 40 times. Below is the sampling distribution of sample proportions of heads. The mean and standard deviation of the sampling distribution are 0.405 and 0.046, respectively. Do you think this was a fair coin? Why or why not?

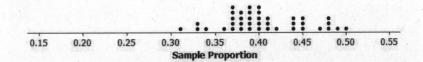

This page intentionally left blank

Lesson 16: Margin of Error When Estimating a Population Proportion

Classwork

Exploratory Challenge 1/Exercises 1–4: Mystery Bag

In this lesson, you will use data from a random sample drawn from a mystery bag to estimate a population proportion and learn how to find and interpret a margin of error for your estimate.

1. Write down your estimate for the proportion of red chips in the mystery bag based on the random sample of 30 chips drawn in class.

2. Tanya and Raoul had a paper bag that contained red and black chips. The bag was marked 40% red chips. They drew random samples of 30 chips, with replacement, from the bag. (They were careful to shake the bag after they replaced a chip.) They had 9 red chips in their sample. They drew another random sample of 30 chips from the bag, and this time they had 12 red chips. They repeated this sampling process 50 times and made a plot of the number of red chips in each sample. A plot of their sampling distribution is shown below.

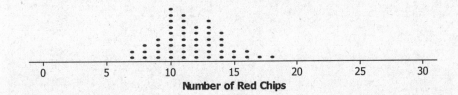

Number of Red Chips

    a. What was the most common number of red chips in the 50 samples? Does this seem reasonable? Why or why not?

b. What number of red chips, if any, never occurred in any of the samples?

c. Give an interval that contains the likely number of red chips in samples of size 30 based on the simulated sampling distribution.

d. Do you think the number of red chips in the mystery bag could have come from a sample drawn from a bag that had 40% red chips? Why or why not?

Nine different bags of chips were distributed to small teams of students in the class. Each bag had a different proportion of red chips. Each team simulated drawing 50 different random samples of size 30 from their bags and recorded the number of red chips for each sample. The graphs of their simulated sampling distributions are shown below.

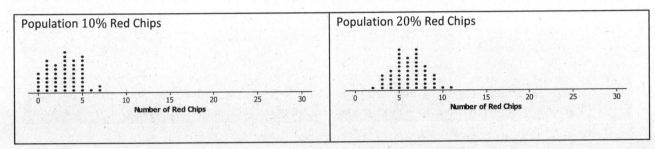

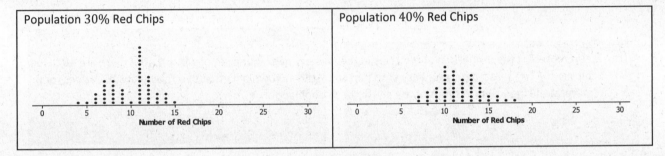

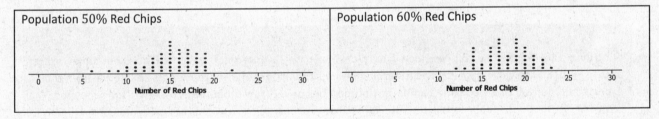

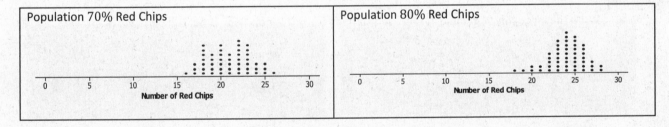

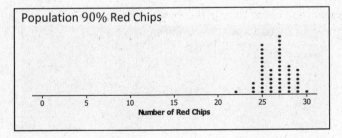

EUREKA MATH™

©2015 Great Minds. eureka-math.org
ALG II-M4-SE-B2-1.3.0-08.2015

3. Think about the number of red chips in the random sample of size 30 that was drawn from the mystery bag.

   a. Based on the simulated sampling distributions, do you think that the mystery bag might have had 10% red chips? Explain your reasoning.

   b. Based on the simulated sampling distributions, which of the percentages 10%, 20%, 30%, 40%, 50%, 60%, 70%, 80%, and 90% might reasonably be the percentage of red chips in the mystery bag?

   c. Let p represent the proportion of red chips in the mystery bag. (For example, $p = 0.40$ if there are 40% red chips in the bag.) Based on your answer to part (b), write an inequality that describes plausible values for p. Interpret the inequality in terms of the mystery bag population.

4. If the inequality like the one you described in part (c) of Exercise 3 went from 0.30 to 0.60, it is sometimes written as 0.45 ± 0.15. The value 0.15 is called a *margin of error*. The margin of error represents an interval from the expected proportion that would not contain any proportions or very few proportions based on the simulated sampling distribution. Proportions in this interval are not expected to occur when taking a sample from the mystery bag.

   a. Write the inequality you found in Exercise 3 part (c), using this notation. What is the margin of error?

   b. Suppose Sol said, "So this means that the actual proportion of red chips in the mystery bag was 60%." Tonya argued that the actual proportion of red chips in the mystery bag was 20%. What would you say?

©2015 Great Minds. eureka-math.org
ALG II-M4-SE-B2-1.3.0-08.2015

Exploratory Challenge 2/Exercises 5–7: Samples of Size 50

5. Do you think the margin of error would be different in Exercise 4 if you had sampled 50 chips instead of 30? Try to convince a partner that your conjecture is correct.

6. Below are simulated sampling distributions of the number of red chips for samples of size 50 from populations with various percentages of red chips.

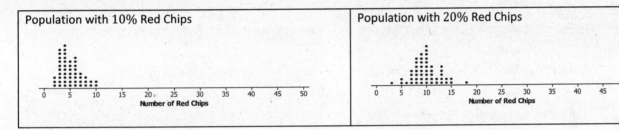

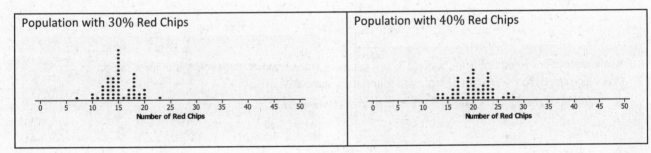

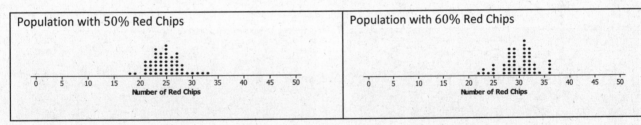

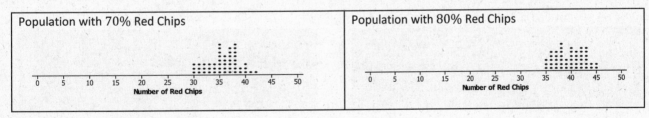

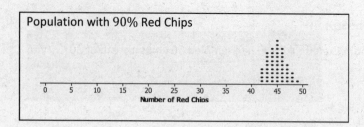

a. Suppose you drew 30 red chips in a random sample of 50 from the mystery bag. What are plausible values for the proportion of red chips in the mystery bag? Explain your reasoning.

b. Write an expression that contains the margin of error based on your answer to part (a).

7. Remember your conjecture from Exercise 5, and compare the margin of error you found for a sample of size 30 (from Exercise 3) to the margin of error you found for a sample of size 50.

a. Was your reasoning in Exercise 5 correct? Why or why not?

b. Explain why the change in the margin of error makes sense.

Lesson 16:   Margin of Error When Estimating a Population Proportion

EUREKA
MATH™

Lesson Summary

In this lesson, you investigated how to make an inference about an unknown population proportion based on a random sample from that population.

- You learned how random samples from populations with known proportions of successes behave by simulating sampling distributions for samples drawn from those populations.

- Comparing an observed proportion of successes from a random sample drawn from a population with an unknown proportion of successes to these sampling distributions gives you some information about what populations might produce a random sample like the one you observed.

- These plausible population proportions can be described as $p \pm M$. The value of M is called a *margin of error*.

Problem Set

1. Tanya simulated drawing a sample of size 30 from a population of chips and got the following simulated sampling distribution for the number of red chips:

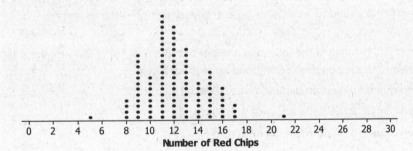

Which of the following results seem like they might have come from this population? Explain your reasoning.

   I.   8 red chips in a random sample of size 30

   II.   12 red chips in a random sample of size 30

   III. 24 red chips in a random sample of size 30

2. 64% percent of the students in a random sample of 100 high school students intended to go to college. The graphs below show the result of simulating random samples of size 100 from several different populations where the success percentage was known and recording the percentage of successes in the sample.

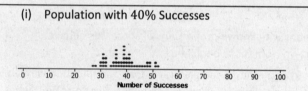

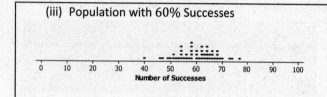

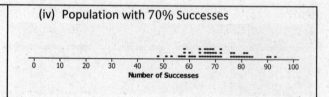

a. Based on these graphs, which of the following are plausible values for the percentage of successes in the population from which the sample was selected: 40%, 50%, 60%, or 70%? Explain your thinking.

b. Would you need more information to determine plausible values for the actual proportion of the population of high school students who intend to go to some postsecondary school? Why or why not?

3. Suppose the mystery bag had resulted in the following number of red chips. Using the simulated sampling distributions found earlier in this lesson, find a margin of error in each case.

a. The number of red chips in a random sample of size 30 was 10.

b. The number of red chips in a random sample of size 30 was 21.

c. The number of red chips in a random sample of size 50 was 22.

4. The following intervals were plausible population proportions for a given sample. Find the margin of error in each case.

a. From 0.35 to 0.65

b. From 0.72 to 0.78

c. From 0.84 to 0.95

d. From 0.47 to 0.57

5. Decide if each of the following statements is true or false. Explain your reasoning in each case.

a. The smaller the sample size, the smaller the margin of error.

b. If the margin of error is 0.05 and the observed proportion of red chips is 0.45, then the true population proportion is likely to be between 0.40 and 0.50.

6. Extension: The margin of error for a sample of size 30 is 0.20; for a sample of 50, it is 0.10. If you increase the sample size to 70, do you think the margin of error for the percent of successes will be 0.05? Why or why not?

Lesson 17: Margin of Error When Estimating a Population Proportion

Classwork

In this lesson, you will find and interpret the standard deviation of a simulated distribution for a sample proportion and use this information to calculate a margin of error for estimating the population proportion.

Exercises 1–6: Standard Deviation for Proportions

In the previous lesson, you used simulated sampling distributions to learn about sampling variability in the sample proportion and the margin of error when using a random sample to estimate a population proportion. However, finding a margin of error using simulation can be cumbersome and take a long time for each situation. Fortunately, given the consistent behavior of the sampling distribution of the sample proportion for random samples, statisticians have developed a formula that will allow you to find the margin of error quickly and without simulation.

1. 30% of students participating in sports at Union High School are female (a proportion of 0.30).

   a. If you took many random samples of 50 students who play sports and made a dot plot of the proportion of female students in each sample, where do you think this distribution will be centered? Explain your thinking.

   b. In general, for any sample size, where do you think the center of a simulated distribution of the sample proportion of female students in sports at Union High School will be?

2. Below are two simulated sampling distributions for the sample proportion of female students in random samples from all the students at Union High School.

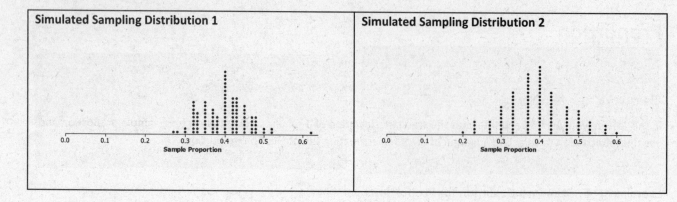

a. Based on the two sampling distributions above, what do you think is the population proportion of female students?

b. One of the sampling distributions above is based on random samples of size 30, and the other is based on random samples of size 60. Which sampling distribution corresponds to the sample size of 30? Explain your choice.

3. Remember from your earlier work in statistics that distributions were described using shape, center, and spread. How was spread measured?

4. For random samples of size n, the standard deviation of the sampling distribution of the sample proportion can be calculated using the following formula:

$$\text{standard deviation} = \sqrt{\frac{p(1-p)}{n}},$$

where p is the value of the population proportion and n is the sample size.

a. If the proportion of female students at Union High School is 0.4, what is the standard deviation of the distribution of the sample proportions of female students for random samples of size 50? Round your answer to three decimal places.

b. The proportion of male students at Union High School is 0.6. What is the standard deviation of the distribution of the sample proportions of male students for samples of size 50? Round your answer to three decimal places.

c. Think about the graphs of the two distributions in parts (a) and (b). Explain the relationship between your answers using the center and spread of the distributions.

5. Think about the simulations that your class performed in the previous lesson and the simulations in Exercise 2 above.

a. Was the sampling variability in the sample proportion greater for samples of size 30 or for samples of size 50? In other words, does the sample proportion tend to vary more from one random sample to another when the sample size is 30 or 50?

b. Explain how the observation that the variability in the sample proportions decreases as the sample size increases is supported by the formula for the standard deviation of the sample proportion.

6. Consider the two simulated sampling distributions of the proportion of female students in Exercise 2 where the population proportion was 0.4. Recall that you found $n = 60$ for Distribution 1 and $n = 30$ for Distribution 2.

   a. Find the standard deviation for each distribution. Round your answer to three decimal places.

   b. Make a sketch, and mark off the intervals one standard deviation from the mean for each of the two distributions. Interpret the intervals in terms of the proportion of female students in a sample.

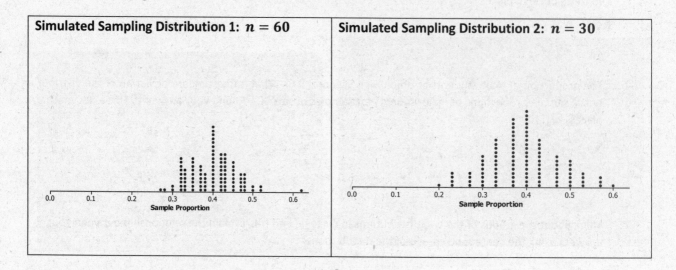

In general, three results about the sampling distribution of the sample proportion are known.

- The sampling distribution of the sample proportion is centered at the actual value of the population proportion, p.

- The sampling distribution of the sample proportion is less variable for larger samples than for smaller samples. The variability in the sampling distribution is described by the standard deviation of the distribution, and the standard deviation of the sampling distribution for random samples of size n is $\sqrt{\dfrac{p(1-p)}{n}}$, where p is the value of the population proportion. This standard deviation is usually estimated using the sample proportion, which is denoted by $\hat{p}$ (read as p-hat), to distinguish it from the population proportion. The formula for the estimated standard deviation of the distribution of sample proportions is $\sqrt{\dfrac{\hat{p}(1-\hat{p})}{n}}$.

- As long as the sample size is large enough that the sample includes at least 10 successes and failures, the sampling distribution is approximately normal in shape. That is, a normal distribution would be a reasonable model for the sampling distribution.

Exercises 7–12: Using the Standard Deviation with Margin of Error

7. In the work above, you investigated a simulated sampling distribution of the proportion of female students in a sample of size 30 drawn from a population with a known proportion of 0.4 female students. The simulated distribution of the proportion of red chips in a sample of size 30 drawn from a population with a known proportion of 0.4 is displayed below.

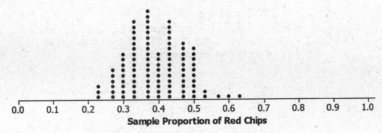

a. Use the formula for the standard deviation of the sample proportion to calculate the standard deviation of the sampling distribution. Round your answer to three decimal places.

b. The distribution from Exercise 2 for a sample of size 30 is below. How do the two distributions compare?

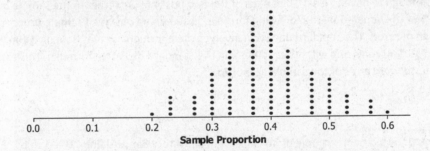

EUREKA
MATH™

Lesson 17:   Margin of Error When Estimating a Population Proportion

S.147

©2015 Great Minds. eureka-math.org
ALG II-M4-SE-B2-1.3.0-08.2015

c. How many of the values of the sample proportions are within one standard deviation of 0.4? How many are within two standard deviations of 0.4?

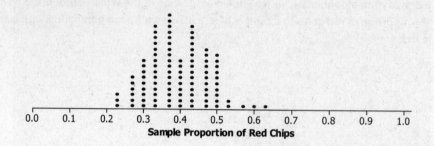

Sample Proportion of Red Chips

In general, for a known population proportion, about 95% of the outcomes of a simulated sampling distribution of a sample proportion will fall within two standard deviations of the population proportion. One caution is that if the proportion is close to 1 or 0, this general rule may not hold unless the sample size is very large. You can build from this to estimate a proportion of successes for an unknown population proportion and calculate a margin of error without having to carry out a simulation.

If the sample is large enough to have at least 10 of each of the two possible outcomes in the sample but small enough to be no more than 10% of the population, the following formula (based on an observed sample proportion $\hat{p}$) can be used to calculate the margin of error. The standard deviation involves the parameter p that is being estimated. Because p is often not known, statisticians replace p with its estimate $\hat{p}$ in the standard deviation formula. This estimated standard deviation is called the *standard error* of the sample proportion.

8.

a. Suppose you draw a random sample of 36 chips from a mystery bag and find 20 red chips. Find $\hat{p}$, the sample proportion of red chips, and the standard error.

b. Interpret the standard error.

Lesson 17: Margin of Error When Estimating a Population Proportion

EUREKA
MATH™

When estimating a population proportion, *margin of error* can be defined as the *maximum expected difference* between the value of the population proportion and a sample estimate of that proportion (the farthest away from the actual population value that you think your estimate is likely to be).

> If $\hat{p}$ is the sample proportion for a random sample of size n from some population and if the sample size is large enough,
>
> $$\text{estimated margin of error} = 2\sqrt{\frac{\hat{p}(1-\hat{p})}{n}}.$$

9. Henri and Terence drew samples of size 50 from a mystery bag. Henri drew 42 red chips, and Terence drew 40 red chips. Find the margins of error for each student.

10. Divide the problems below among your group, and find the sample proportion of successes and the estimated margin of error in each situation:

   a. Sample of size 20, 5 red chips

   b. Sample of size 40, 10 red chips

   c. Sample of size 80, 20 red chips

   d. Sample of size 100, 25 red chips

11. Look at your answers to Exercise 2.

    a. What conjecture can you make about the relation between sample size and margin of error? Explain why your conjecture makes sense.

    b. Think about the formula for a margin of error. How does this support or refute your conjecture?

12. Suppose that a random sample of size 100 will be used to estimate a population proportion.

    a. Would the estimated margin of error be greater if $\hat{p} = 0.4$ or $\hat{p} = 0.5$? Support your answer with appropriate calculations.

    b. Would the estimated margin of error be greater if $\hat{p} = 0.5$ or $\hat{p} = 0.8$? Support your answer with appropriate calculations.

    c. For what value of $\hat{p}$ do you think the estimated margin of error will be greatest? (Hint: Draw a graph of $\hat{p}(1 - \hat{p})$ as $\hat{p}$ ranges from 0 to 1.)

Lesson Summary

- Because random samples behave in a consistent way, a large enough sample size allows you to find a formula for the standard deviation of the sampling distribution of a sample proportion. This can be used to calculate the margin of error: $M = 2\sqrt{\dfrac{\hat{p}(1-\hat{p})}{n}}$, where $\hat{p}$ is the proportion of successes in a random sample of size n.

- The sample size is large enough to use this result for estimated margin of error if there are at least 10 of each of the two outcomes.

- The sample size should not exceed 10% of the population.

- As the sample size increases, the margin of error decreases.

Problem Set

1. Different students drew random samples of size 50 from the mystery bag. The number of red chips each drew is given below. In each case, find the margin of error for the proportions of the red chips in the mystery bag.

   a. 10 red chips

   b. 28 red chips

   c. 40 red chips

2. The school newspaper at a large high school reported that 120 out of 200 randomly selected students favor assigned parking spaces. Compute the margin of error. Interpret the resulting interval in context.

3. A newspaper in a large city asked 500 women the following: "Do you use organic food products (such as milk, meats, vegetables, etc.)?" 280 women answered "yes." Compute the margin of error. Interpret the resulting interval in context.

4. The results of testing a new drug on 1,000 people with a certain disease found that 510 of them improved when they used the drug. Assume these 1,000 people can be regarded as a random sample from the population of all people with this disease. Based on these results, would it be reasonable to think that more than half of the people with this disease would improve if they used the new drug? Why or why not?

5. A newspaper in New York took a random sample of 500 registered voters from New York City and found that 300 favored a certain candidate for governor of the state. A second newspaper polled 1,000 registered voters in upstate New York and found that 550 people favored this candidate. Explain how you would interpret the results.

6. In a random sample of 1,500 students in a large suburban school, 1,125 reported having a pet, resulting in the interval 0.75 ± 0.022. In a large urban school, 840 out of 1,200 students reported having a pet, resulting in the interval 0.7 ± 0.026. Because these two intervals do not overlap, there appears to be a difference in the proportion of suburban students owning a pet and the proportion of urban students owning a pet. Suppose the sample size of the suburban school was only 500, but 75% still reported having a pet. Also, suppose the sample size of the urban school was 600, and 70% still reported having a pet. Is there still a difference in the proportion of students owning a pet in suburban schools and urban schools? Why does this occur?

7. Find an article in the media that uses a margin of error. Describe the situation (an experiment, an observational study, a survey), and interpret the margin of error for the context.

EUREKA
MATH™

Lesson 18: Sampling Variability in the Sample Mean

Classwork

Exploratory Challenge/Exercises 1–7: Random Segments

The worksheet contains 100 segments of different lengths. The length of a segment is the number of rectangles spanned on the grid. For example, segment 2 has length 5.

1. Briefly review the sheet, and estimate the mean length of the segments. Will your estimate be close to the actual mean? Why or why not?

2. Look at the sheet. With which of the statements below would you agree? Explain your reasoning.

    The mean length of the segments is

    a. Close to 1.

    b. Close to 8.

    c. Around 5.

    d. Between 2 and 5.

3. Follow your teacher's directions to select ten random numbers between 1 and 100. For each random number, start at the upper left cell with a segment value of 2, and count down and to the right the number of cells based on the random number selected. The number in the cell represents the length of a randomly selected segment.

    a. On a number line, graph the lengths of the corresponding segments on the worksheet.

b.   Find the mean and standard deviation of the lengths of the segments in your sample. Mark the mean length on your graph from part (a).

4.   Your sample provides some information about the mean length of the segments in one random sample of size 10, but that sample is only one among all the different possible random samples. Let's look at other random samples and see how the means from those samples compare to the mean segment length from your random sample.

Record the mean segment length for your random sample on a Post-it note, and post the note in the appropriate place on the number line your teacher set up.

a.   Jonah looked at the plot and said, "Wow. Our means really varied." What do you think he meant?

b.   Describe the simulated sampling distribution of mean segment lengths for samples of size 10.

c.   How did your first estimate (from Exercise 1) compare to your sample mean from the random sample? How did it compare to the means in the simulated distribution of the sample means from the class?

5.   Collect the values of the sample means from the class.

a.   Find the mean and standard deviation of the simulated distribution of the sample means.

b.   Interpret the standard deviation of the simulated sampling distribution in terms of the length of the segments.

©2015 Great Minds. eureka-math.org
ALG II-M4-SE-B2-1.3.0-08.2015

c.   What do you observe about the values of the means in the simulated sampling distribution that are within two standard deviations from the mean of the sampling distribution?

6.   Generate another set of ten random numbers, find the corresponding lengths on the sheet, and calculate the mean length for your sample. Put a Post-it note with your sample mean on the second number line. Then, answer the following questions.

a.   Find the mean and standard deviation of the simulated distribution of the sample means.

b.   Interpret the standard deviation of the simulated sampling distribution in terms of the length of the segments.

c.   What do you observe about the values of the means in the simulated sampling distribution that are within two standard deviations from the mean of the sampling distribution?

7.   Suppose that we know the actual mean of all the segment lengths is 2.78 units.

a.   Describe how the population mean relates to the two simulated distributions of sample means.

b.   Tonya was concerned that neither of the simulated distributions of sample means had a value around 5, but some of the segments on the worksheet were 5 units long, and some were as big as 8 units long. What would you say to Tonya?

Lesson Summary

In this lesson, you drew a sample from a population and found the mean of that sample.

- Drawing many samples of the same size from the same population and finding the mean of each of those samples allows you to build a simulated sampling distribution of the sample means for the samples you generated.

- The mean of the simulated sampling distribution of sample means is close to the population mean.

- In the two examples of simulated distributions of sample means we generated, most of the sample means seemed to fall within two standard deviations of the mean of the simulated distribution of sample means.

Problem Set

1. The three distributions below relate to the population of all of the random segment lengths and to samples drawn from that population. The eight phrases below could be used to describe a whole graph or a value on the graph. Identify where on the appropriate graph the phrases could be placed. (For example, *segment of length* 2 could be placed by any of the values in the column for 2 on the plot labeled *Length.*)

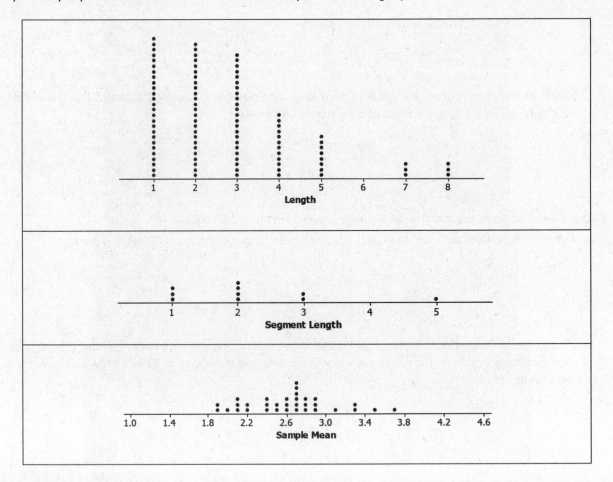

EUREKA
MATH™

a. Random sample of size 10 of segment lengths

b. Segment of length 2

c. Sample mean segment length of 2

d. Mean of sampling distribution, 2.6

e. Simulated distribution of sample means

f. Sample segment lengths

g. Population of segment lengths

h. Mean of all segment lengths, 2.78

2. The following segment lengths were selected in four different random samples of size 10.

Lengths Sample A	Lengths Sample B	Lengths Sample C	Lengths Sample D
1	1	1	2
2	3	5	2
1	1	1	7
5	2	3	2
3	1	4	5
1	5	2	2
2	3	2	3
2	4	4	5
3	3	3	5
1	3	4	4

a. Find the mean segment length of each sample.

b. Find the mean and standard deviation of the four sample means.

c. Interpret your answer to part (b) in terms of the variability in the sampling process.

3. Two simulated sampling distributions of the mean segment lengths from random samples of size 10 are displayed below.

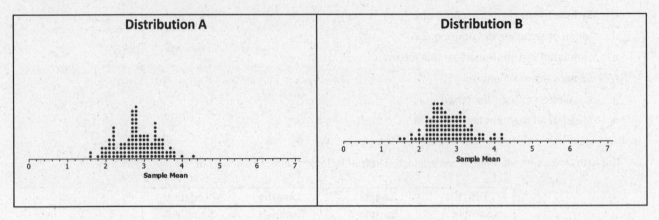

a. Compare the two distributions with respect to shape, center, and spread.

b. Distribution A has a mean of 2.82, and Distribution B has a mean of 2.77. How do these means compare to the population mean of 2.78?

c. Both Distribution A and Distribution B have a standard deviation 0.54. Make a statement about the distribution of sample means that makes use of this standard deviation.

4. The population distribution of all the segment lengths is shown in the dot plot below. How does the population distribution compare to the two simulated sampling distributions of the sample means in Problem 3?

Distribution of Lengths of 100 Segments

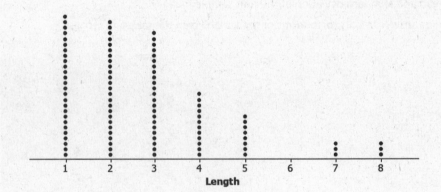

EUREKA
MATH

Exercises 1–7: Random Segments

EUREKA
MATH

Lesson 18:   Sampling Variability in the Sample Mean

S.159

©2015 Great Minds. eureka-math.org
ALG II-M4-SE-B2-1.3.0-08.2015

This page intentionally left blank

Lesson 19: Sampling Variability in the Sample Mean

Classwork

This lesson uses simulation to approximate the sampling distribution of the sample mean for random samples from a population, explores how the simulated sampling distribution provides insight into the anticipated estimation error when using a sample mean to estimate a population mean, and covers how sample size affects the distribution of the sample mean.

Exercises 1–6: SAT Scores

1. SAT scores vary a lot. The table on the next page displays the 506 scores for students in one New York school district for a given year.

   a. Looking at the table, how would you describe the population of SAT scores?

   b. Jason used technology to draw a random sample of size 20 from all of the scores and found a sample mean of 487. What does this value represent in terms of the graph below?

Random Sample from District SAT Scores

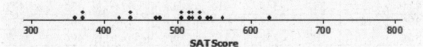

SAT Score

Table 1: SAT Scores for District Students

441	395	369	350	521	691	648	521	498	413	486
440	415	481	392	800	448	603	503	486	476	500
391	359	447	550	432	158	379	394	495	442	507
395	504	399	424	456	729	356	392	514	388	518
445	436	386	493	467	493	440	387	512	431	467
499	412	457	389	323	319	550	450	517	405	506
486	519	369	373	348	532	496	488	504	444	
396	473	319	367	679	472	613	561	522	408	
451	427	369	560	602	520	567	495	473	424	
362	391	371	407	436	366	582	528	533	463	
328	613	357	438	436	713	603	525	553	446	
414	466	382	362	777	259	557	508	495	466	
409	486	627	589	749	410	639	516	520	632	
526	334	608	374	634	443	556	506	506	526	
391	497	378	358	566	442	496	568	544	546	
529	392	387	373	198	555	499	476	525	529	
529	426	470	378	345	431	613	490	548	455	
574	379	380	561	712	197	556	547	543	431	
363	382	370	379	504	254	596	489	474	386	
486	434	365	530	685	372	580	506	529	434	
418	722	674	504	645	501	605	511	566	362	
527	437	388	525	509	662	445	489	487	426	
441	395	377	561	448	503	602	523	510	404	
467	463	427	519	491	448	638	530	518	493	
387	433	446	525	352	662	570	507	515	515	
503	371	394	569	779	158	558	504	516	407	
350	392	368	484	689	691	535	522	505	409	
583	416	406	416	513	729	623	503	536	422	
370	370	350	446	624	493	465	524	547	612	
499	422	344	420	465	319	460	523	528	486	
399	532	347	446	504	532	375	524	527	394	
374	545	377	462	390	472	540	501	523	424	
372	427	391	528	576	520	564	482	540	393	
559	371	339	533	756	366	547	502	480	420	
330	390	404	543	451	713	568	503	516	415	
567	529	377	460	505	259	588	439	501	394	
371	341	469	391	540	410	502	474	452	473	
503	356	417	623	436	443	510	477	507	531	
327	351	356	587	298	442	589	458	486	469	
528	377	370	528	449	555	537	494	500	453	
447	404	355	356	352	431	410	447	507	442	
572	369	364	523	574	197	330	517	518	509	
379	396	383	404	518	460	500	457	467	435	
456	396	400	505	682	623	531	471	506	427	
406	535	404	512	474	587	509	541	509	489	
420	388	375	514	629	528	571	513	597	480	
395	370	398	516	656	523	527	441	509	516	
355	417	376	498	539	505	457	489	567	501	
423	419	451	460	553	514	552	498	509	452	
438	348	369	541	400	629	561	538	597	507	

Lesson 19:   Sampling Variability in the Sample Mean

EUREKA MATH™

2. If you were to take many different random samples of 20 from this population, describe what you think the sampling distribution of these sample means would look like.

3. Everyone in Jason's class drew several random samples of size 20 and found the mean SAT score. The plot below displays the distribution of the mean SAT scores for their samples.

Random Sample from District SAT Scores

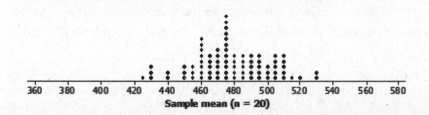

a. How does the simulated sampling distribution compare to your conjecture in Exercise 2? Explain any differences.

b. Use technology to generate many more samples of size 20, and plot the means of those samples. Describe the shape of the simulated distribution of sample mean SAT scores.

c. How did the simulated distribution using more samples compare to the one you generated in Exercise 3?

d. What are the mean and standard deviation of the simulated distribution of the sample mean SAT scores you found in part (b)? (Use technology and your simulated distribution of the sample means to find the values.)

e. Write a sentence describing the distribution of sample means that uses the mean and standard deviation you calculated in part (d).

4. Reflect on some of the simulated sampling distributions you have considered in previous lessons.

a. Make a conjecture about how you think the size of the sample might affect the distribution of the sample SAT means.

b. To test the conjecture, investigate the sample sizes 5, 10, 40, 50, and the simulated distribution of sample means from Exercise 3. Divide the sample sizes among your group members, and use technology to simulate sampling distributions of mean SAT scores for samples of the different sizes. Find the mean and standard deviation of each simulated sampling distribution.

c. How does the sample size seem to affect the simulated distributions of the sample SAT mean scores? Include the simulated distribution from part (b) of Exercise 3 in your response. Why do you think this is true?

©2015 Great Minds. eureka-math.org
ALG II-M4-SE-B2-1.3.0-08.2015

5.

    a.   For each of the sample sizes, consider how the standard deviation seems to be related to the range of the sample means in the simulated distributions of the sample SAT means you found in Exercise 4.

    b.   How do your answers to part (a) compare to the answers from other groups?

6.

    a.   Make a graph of the distribution of the population consisting of the SAT scores for all of the students.

    b.   Find the mean of the distribution of SAT scores. How does it compare to the mean of the sampling distributions you have been simulating?

©2015 Great Minds. eureka-math.org
ALG II-M4-SE-B2-1.3.0-08.2015

Lesson Summary

For a given sample, you can find the sample mean.

- There is variability in the sample mean. The value of the sample mean varies from one random sample to another.

- A graph of the distribution of sample means from many different random samples is a simulated sampling distribution.

- Sample means from random samples tend to cluster around the value of the population mean. That is, the simulated sampling distribution of the sample mean will be centered close to the value of the population mean.

- The variability in the sample mean decreases as the sample size increases.

- Most sample means are within two standard deviations of the mean of the simulated sampling distribution.

Problem Set

1. Which of the following will have the smallest standard deviation? Explain your reasoning.

   A sampling distribution of sample means for samples of size:

   a. 15   b. 25   c. 100

2. In light of the distributions of sample means you have investigated in the lesson, comment on the statements below for random samples of size 20 chosen from the district SAT scores.

   a. Josh claimed he took a random sample of size 20 and had a sample mean score of 320.

   b. Sarfina stated she took a random sample of size 20 and had a sample mean of 520.

   c. Ana announced that it would be pretty rare for the mean SAT score in a random sample to be more than three standard deviations from the mean SAT score of 475.

3. Refer to your answers for Exercise 4, and then comment on each of the following:

   a. A random sample of size 50 produced a mean SAT score of 400.

   b. A random sample of size 10 produced a mean SAT score of 400.

   c. For what sample sizes was a sample mean SAT score of 420 plausible? Explain your thinking.

4. Explain the difference between the sample mean and the mean of the sampling distribution.

Lesson 20: Margin of Error When Estimating a Population Mean

Classwork

Example 1: Describing a Population of Numerical Data

The course project in a computer science class was to create 100 computer games of various levels of difficulty that had ratings on a scale from 1 (easy) to 20 (difficult). We will examine a representation of the data resulting from this project. Working in pairs, your teacher will give you a page that contains 100 rectangles of various sizes.

a. What do you think the rectangles represent in the context of the 100 computer games?

b. What do you think the sizes of the rectangles represent in the context of the 100 computer games?

c. Why do you think the rectangles are numbered from 00 to 99 instead of from 1 to 100?

Exploratory Challenge 1/Exercises 1–3: Estimate the Population Mean Rating

1. Working with your partner, discuss how you would calculate the mean rating of all 100 computer games (the population mean).

2. Discuss how you might select a random sample to estimate the population mean rating of all 100 computer games.

©2015 Great Minds. eureka-math.org
ALG II-M4-SE-B2-1.3.0-08.2015

3. Calculate an estimate of the population mean rating of all 100 computer games based on a random sample of size 10. Your estimate is called a *sample mean,* and it is denoted by $\bar{x}$. Use the following random numbers to select your sample.

34 86 80 58 04 43 96 29 44 51

Exploratory Challenge 2/Exercises 4–6: Build a Distribution of Sample Means

4. Work in pairs. Using a table of random digits or a calculator with a random number generator, generate four sets of ten random numbers. Use these sets of random numbers to identify four random samples of size 10. Calculate the sample mean rating for each of your four random samples.

5. Write your sample means on separate sticky notes, and post them on a number line that your teacher has prepared for your class.

6. The actual population mean rating of all 100 computer games is 7.5. Does your class distribution of sample means center at 7.5? Discuss why it does or does not.

Example 2: Margin of Error

Suppose that 50 random samples, each of size ten, produced the sample means displayed in the following dot plot:

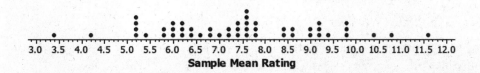

Sample Mean Rating

Note that almost all of the sample means are between 4 and 11. That is, almost all are roughly within 3.5 rating points of the population mean 7.5. The value 3.5 is a visual estimate of the margin of error. It is not really an "error" in the sense of a "mistake." Rather, it is how far our estimate for the population mean is likely to be from the actual value of the population mean.

Based on the class distribution of sample means, is the visual estimate of margin of error close to 3.5?

Example 3: Standard Deviation as a Refinement of Margin of Error

Note that the margin of error is measuring how spread out the sample means are relative to the value of the actual population mean. From previous lessons, you know that the standard deviation is a good measure of spread. So, rather than producing a visual estimate for the margin of error from the distribution of sample means, another approach is to use the standard deviation of the sample means as a measure of spread. For example, the standard deviation of the 50 sample means in the example above is 1.7. Note that if you double 1.7, you get a value for margin of error close to the visual estimate of 3.5.

Another way to estimate margin of error is to use two times the standard deviation of a distribution of sample means. For the above example, because $2(1.7) = 3.4$, the refined margin of error (based on the standard deviation of sample means) is 3.4 rating points.

An interpretation of the margin of error is that plausible values for the population mean rating are from $7.5 - 3.4$ to $7.5 + 3.4$ (i.e., 4.1 to 10.9 rating points).

Exploratory Challenge 3/Exercise 7

7.   Calculate and interpret the margin of error for your estimate of the population mean rating of 100 computer games based on the standard deviation of your class distribution of sample means.

Lesson Summary

This lesson revisited margin of error. Previously, you estimated a population proportion of successes and described the accuracy of the estimate by its margin of error. This lesson also focused on margin of error but in the context of estimating the mean of a population of numerical data.

Margin of error was estimated in two ways.

- The first was through a visual estimation in which you judged the amount of spread in the distribution of sample means.

- The second was more formalized by defining margin of error as twice the standard deviation of the distribution of sample means.

Problem Set

1. Suppose you are interested in knowing how many text messages eleventh graders send daily.

   Describe the steps that you would take to estimate the mean number of text messages per day sent by all eleventh graders at a school.

2. Suppose that 62 random samples based on ten student responses to the question, "How many text messages do you send per day?" resulted in the 62 sample means (rounded) shown below.

65	68	76	76	78	82	83	83	85	86	87	88	88
88	89	89	89	90	91	91	91	91	92	92	92	92
92	93	93	93	93	93	94	94	94	94	94	94	95
95	95	95	95	95	95	95	96	96	97	97	97	98
98	98	98	98	99	100	100	101	104	106			

   a. Draw a dot plot for the distribution of sample means.

   b. Based on your dot plot, would you be surprised if the actual mean number of text messages sent per day for all eleventh graders in the school is 91.7? Why or why not?

3. Determine a visual estimate of the margin of error when a random sample of size 10 is used to estimate the population mean number of text messages sent per day.

4. The standard deviation of the above distribution of sample mean number of text messages sent per day is 7.5. Use this to calculate and interpret the margin of error for an estimate of the population mean number of text messages sent daily by eleventh graders (based on a random sample of size 10 from this population).

Example 1: Describing a Population of Numerical Data

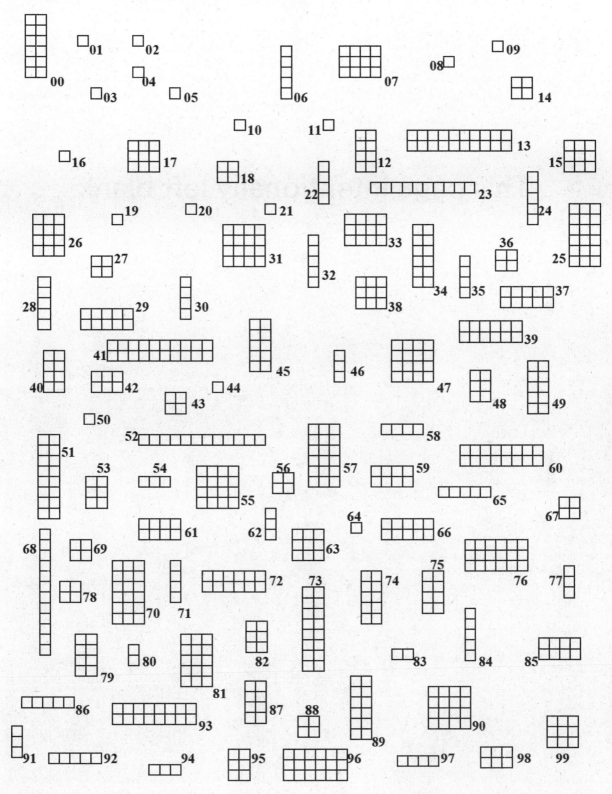

EUREKA MATH™

©2015 Great Minds. eureka-math.org
ALG II-M4-SE-B2-1.3.0-08.2015

This page intentionally left blank

Lesson 21: Margin of Error When Estimating a Population Mean

Classwork

This lesson continues to discuss using the sample mean as an estimate of the population mean and judging its accuracy based on the concept of margin of error. In the last lesson, the margin of error was defined as twice the standard deviation of the sampling distribution of the sample mean. In this lesson, a formula will be given for the margin of error that allows you to calculate the margin of error from a single random sample rather than having to create a sampling distribution of sample means.

Example 1: Estimating a Population Mean Using a Random Sample

Provide a one-sentence summary of our findings from the previous lesson.

What were drawbacks of the calculation method?

In practice, you do not have to use that process to find the margin of error. Fortunately, just as was the case with estimating a population proportion, there are some general results that lead to a formula that allows you to estimate the margin of error using a single sample. You can then gauge the accuracy of your estimate of the population mean by calculating the margin of error using the sample standard deviation.

> The standard deviation of the distribution of sample means is approximated by $\frac{s}{\sqrt{n}}$, where s is the standard deviation of the sample and n is the size of the sample.

Exercises 1–5

1. Suppose a random sample of size ten produced the following ratings in the computer games rating example in the last lesson: $12, 5, 2, 4, 1, 4, 18, 10, 1, 16$. Estimate the population mean rating based on these ten sampled ratings.

2. Calculate the sample standard deviation. Round your answer to three decimal places.

3. Use the formula given above to calculate the approximate standard deviation of the distribution of sample means. Round your answer to three decimal places.

4. Recall that the margin of error is twice the standard deviation of the distribution of sample means. What is the value of the margin of error based on this sample? Write a sentence interpreting the value of the margin of error in the context of this problem on computer game ratings.

5. Based on the sample mean and the value of the margin of error, what is an interval of plausible values for the population mean?

Exercises 6–13: The Gettysburg Address

The Gettysburg Address is considered one of history's greatest speeches. Some students noticed that the speech was very short (about 268 words, depending on the version) and wondered if the words were also relatively short. To estimate the mean length of words in the population of words in the Gettysburg Address, work with a partner on the following steps. Your teacher will give you a copy of the Gettysburg Address with words numbered from 001 to 268.

6. Develop and describe a plan for collecting data from the Gettysburg Address and determining the typical length of a word. Then, implement your plan, and report your findings.

7. Use a random number table or a calculator with a random number generator to obtain ten different random numbers from 001 to 268.

8. Use the random numbers found in Exercise 7 as identification numbers for the words that will make up your random sample of words from the Gettysburg Address. Make a list of the ten words in your sample.

9. Count the number of letters in each of the ten words in your sample.

10. Calculate the sample mean number of letters for the ten words in your sample.

11. Calculate the sample standard deviation of the number of letters for the ten words in your sample. Round your answer to three decimal places.

12. Use the sample standard deviation from Exercise 11 to calculate the margin of error associated with using your sample mean as an estimate of the population mean. Round your answer to three decimal places.

13. Write a few sentences describing what you have learned about the mean length of the population of 268 words in the Gettysburg Address. Be sure to include an interpretation of the margin of error.

Lesson Summary

- When using the sample mean to estimate a population mean, it is important to know something about how accurate that estimate might be.

- Accuracy can be described by the margin of error.

- The margin of error can be estimated using data from a single random sample (without the need to create a simulated sampling distribution) by using the formula $2\left(\frac{s}{\sqrt{n}}\right)$, where s is the standard deviation of a single sample and n is the sample size.

Problem Set

1. A new brand of hot dog claims to have a lower sodium content than the leading brand.

   a. A random sample of ten of these new hot dogs results in the following sodium measurements (in milligrams).

      370 326 322 297 326 289 293 264 327 331

      Estimate the population mean sodium content of this new brand of hot dog based on the ten sampled measurements.

   b. Calculate the margin of error associated with your estimate of the population mean from part (a). Round your answer to three decimal places.

   c. The mean sodium content of the leading brand of hot dogs is known to be 350 mg. Based on the sample mean and the value of the margin of error for the new brand, is a mean sodium content of 350 mg a plausible value for the mean sodium content of the new brand? Comment on whether you think the new brand of hot dog has a lower sodium content on average than the leading brand.

   d. Another random sample of 40 new-brand hot dogs is taken. Should this larger sample of hot dogs produce a more accurate estimate of the population mean sodium content than the sample of size 10? Explain your answer by appealing to the formula for margin of error.

2. It is well known that astronauts increase their height in space missions because of the lack of gravity. A question is whether or not we increase height here on Earth when we are put into a situation where the effect of gravity is minimized. In particular, do people grow taller when confined to a bed? A study was done in which the heights of six men were taken before and after they were confined to bed for three full days.

   a. The before-after differences in height measurements (in millimeters) for the six men were

      12.6 14.4 14.7 14.5 15.2 13.5.

      Assuming that the men in this study are representative of the population of all men, what is an estimate of the population mean increase in height after three full days in bed?

   b. Calculate the margin of error associated with your estimate of the population mean from part (a). Round your answer to three decimal places.

   c. Based on your sample mean and the margin of error from parts (a) and (b), what are plausible values for the population mean height increase for all men who stay in bed for three full days?

This page intentionally left blank

Exercises 6–13: The Gettysburg Address

001 Four	045 any	089 nation	133 our	177 they	221 full	265 perish
002 score	046 nation,	090 might	134 poor	178 who	222 measure	266 from
003 and	047 so	091 live.	135 power	179 fought	223 of	267 the
004 seven	048 conceived	092 It	136 to	180 here	224 devotion,	268 earth.
005 years	049 and	093 is	137 add	181 have	225 that	
006 ago,	050 so	094 altogether	138 or	182 thus	226 we	
007 our	051 dedicated,	095 fitting	139 detract.	183 far	227 here	
008 fathers	052 can	096 and	140 The	184 so	228 highly	
009 brought	053 long	097 proper	141 world	185 nobly	229 resolve	
010 forth	054 endure.	098 that	142 will	186 advanced.	230 that	
011 upon	055 We	099 we	143 little	187 It	231 these	
012 this	056 are	100 should	144 note,	188 is	232 dead	
013 continent	057 met	101 do	145 nor	189 rather	233 shall	
014 a	058 on	102 this.	146 long	190 for	234 not	
015 new	059 a	103 But,	147 remember,	191 us	235 have	
016 nation;	060 great	104 in	148 what	192 to	236 died	
017 conceived	061 battlefield	105 a	149 we	193 be	237 in	
018 in	062 of	106 larger	150 say	194 here	238 vain,	
019 liberty,	063 that	107 sense,	151 here,	195 dedicated	239 that	
020 and	064 war.	108 we	152 but	196 to	240 this	
021 dedicated	065 We	109 cannot	153 it	197 the	241 nation,	
022 to	066 have	110 dedicate,	154 can	198 great	242 under	
023 the	067 come	111 we	155 never	199 task	243 God,	
024 proposition	068 to	112 cannot	156 forget	200 remaining	244 shall	
025 that	069 dedicate	113 consecrate,	157 what	201 before	245 have	
026 all	070 a	114 we	158 they	202 us,	246 a	
027 men	071 portion	115 cannot	159 did	203 that	247 new	
028 are	072 of	116 hallow	160 here.	204 from	248 birth	
029 created	073 that	117 this	161 It	205 these	249 of	
030 equal.	074 field	118 ground.	162 is	206 honored	250 freedom,	
031 Now	075 as	119 The	163 for	207 dead	251 and	
032 we	076 a	120 brave	164 us	208 we	252 that	
033 are	077 final	121 men,	165 the	209 take	253 government	
034 engaged	078 resting	122 living	166 living,	210 increased	254 of	
035 in	079 place	123 and	167 rather,	211 devotion	255 the	
036 a	080 for	124 dead,	168 to	212 to	256 people,	
037 great	081 those	125 who	169 be	213 that	257 by	
038 civil	082 who	126 struggled	170 dedicated	214 cause	258 the	
039 war,	083 here	127 here	171 here	215 for	259 people,	
040 testing	084 gave	128 have	172 to	216 which	260 for	
041 whether	085 their	129 consecrated	173 the	217 they	261 the	
042 that	086 lives	130 it,	174 unfinished	218 gave	262 people,	
043 nation,	087 that	131 far	175 work	219 the	263 shall	
044 or	088 that	132 above	176 which	220 last	264 not	

This page intentionally left blank

Lesson 22: Evaluating Reports Based on Data from a Sample

Classwork

Exercises 1–5: Election Results

The following is part of an article that appeared in a newspaper:

> With the election for governor still more than a year away, a new poll shows the race is already close. The Republican governor had 47%, and the Democratic challenger had 45% in a poll released Tuesday of 800 registered voters.

> "That's within the poll's margin of error of 3.5 percentage points, making it essentially a toss-up," said the poll's director.

1. Why don't the two percentages add up to 100%?

2. What is meant by the margin of error of 3.5 percentage points?

3. Using the sample size of 800 and the proportion 0.47, calculate the margin of error associated with the estimate of the proportion of all registered voters who would vote for the Republican governor.

4. Why did the poll director say that the election is "essentially a toss-up"?

5. If the sample size had been 2,500 registered voters, and the results stated 47% would vote for the Republican governor, and 45% said they would vote for the Democratic challenger, what would the margin of error have been? Could the director still say that the election was a toss-up?

Exercises 6–8: Chocolate Chip Claim

The Nabisco Company claims that there are at least 1,000 chocolate chips in every 18-ounce bag of their Chips Ahoy! cookies. An article in a local newspaper reported the efforts of a group of students in their attempt to validate the Nabisco claim. The article reported that the students randomly selected 42 bags of cookies from local grocery stores and counted the number of chocolate chips in the cookies in each bag. The students found the sample mean was 1,261.6 chips, and the sample standard deviation was 117.6 chips. The article stated that the students' data supported the Nabisco Company claim.

6. Using the students' results, calculate the margin of error associated with the estimate of the mean number of chocolate chips in an 18-ounce bag of Chips Ahoy! chocolate chip cookies. Write a sentence interpreting the margin of error.

7. Do you agree that the student data supported the Nabisco Company claim? Explain.

8. Comment on the procedure that the students used to collect their data.

©2015 Great Minds. eureka-math.org
ALG II-M4-SE-B2-1.3.0-08.2015

Exercises 9–15: Understanding a Poll

George Gallup founded the American Institute of Public Opinion (Gallup Poll) in 1935. The company is famous for its public opinion polls, which are conducted in the United States and other countries.

Gallup published a graph in May 2013 titled *Percent in U.S. Who Exercise for at Least 30 Minutes Three or More Days per Week*. Use the graph found on the Gallup website (http://www.gallup.com/poll/162194/americans-exercise-habits-worsen-slightly-2013.aspx) to answer the following questions:

9. What percent of those surveyed said that they exercise at least 30 minutes three or more days a week at the start of 2013?

10. Describe the patterns that you observe in the graph.

11. Give some reasons why you think the graph follows the pattern that you described.

Following are the survey methods that Gallup used to collect the data:

> Results are based on telephone interviews conducted as part of the Gallup-Healthways Well-Being Index survey June 1–30, 2013, with a random sample of 15,235 adults, aged 18 and older, living in all 50 U.S. states and the District of Columbia.

> For results based on the total sample of national adults, one can say with 95% confidence that the maximum margin of sampling error is ±1 percentage point.

12. Using the value of 0.538 for the proportion of those surveyed who said they exercise at least 30 minutes three or more days a week in the most recent poll, calculate the margin of error. How does your margin of error compare to the value reported by Gallup?

13. Interpret the phrase "margin of sampling error is ±1 percentage point."

14. Why is it important that Gallup selects a random sample of adults?

15. If Gallup had used a random sample of 1,500, what would happen to the margin of error? Explain your answer.

Lesson Summary

- The estimated margin of error when a sample proportion from a random sample is used to estimate a population proportion is $ME = 2\sqrt{\dfrac{\hat{p}(1-\hat{p})}{n}}$, where $\hat{p}$ is the sample proportion.

- The estimated margin of error when a sample mean from a random sample is used to estimate a population mean is $ME = 2\left(\dfrac{s}{\sqrt{n}}\right)$, where $\bar{x}$ is the sample mean.

- It is important to interpret margin of error in context.

- It is unlikely that the estimate of a population proportion or mean will be farther from the actual population value than the margin of error.

Problem Set

1. The *British Medical Journal* published a study whose objective was to investigate estimation of calorie content of meals from fast food restaurants. Below are the published results.

   Participants: 1,877 adults and 330 school-age children visiting restaurants at dinnertime (evening meal) in 2010 and 2011; 1,178 adolescents visiting restaurants after school or at lunchtime in 2010 and 2011

   Results: Among adults, adolescents, and school-age children, the mean actual calorie content of meals was 836 calories (SD 465), 756 calories (SD 455), and 733 calories (SD 359), respectively. Compared with the actual figures, participants underestimated calorie content by means of 175 calories, 259 calories, and 175 calories, respectively.

   Source: http://www.bmj.com/content/346/bmj.f2907

   a. Calculate the margin of error associated with the estimate of the mean number of actual calories in the meals eaten by each of the groups: adults, adolescents, and school-age children.

   b. Write a sentence interpreting the margin of error for the adult group.

   c. Explain why the margin of error for the estimate of the mean number of actual calories in meals eaten by adults is smaller than the margin of error of the mean number of actual calories in meals eaten by school-age children.

   d. Write a conclusion that the researchers could draw from this study.

©2015 Great Minds. eureka-math.org
ALG II-M4-SE-B2-1.3.0-08.2015

2. The Gallup organization published the following results from a poll that it conducted.

    By their own admission, many young Americans, aged 18 to 29, say they spend too much time using the Internet (59%), their cell phones or smartphones (58%), and social media sites such as Facebook (48%). Americans' perceptions that they spend "too much" time using each of these technologies decline with age. Conversely, older Americans are most likely to say they spend too much time watching television, and among all Americans, television is the most overused technology tested.

    Results are based on telephone interviews conducted as part of Gallup Daily tracking April 9–10, 2012, with a random sample of 1,051 adults, aged 18 and older, living in all 50 U.S. states and the District of Columbia.

    For results based on the total sample of national adults, one can say with 95% confidence that the maximum margin of sampling error is ±4 percentage points.

    Source: http://www.gallup.com/poll/153863/Young-Adults-Admit-Time-Cell-Phones-Web.aspx

    a. Write a newspaper headline that would capture the main idea from the poll.

    b. Use the phrase from the article, "their cell phones or smartphones (58%)," to calculate the margin of error. Show your work.

    c. How do your results compare with the margin of error stated in the article?

    d. Interpret the statement "the margin of sampling error is ±3 percentage points."

    e. What would happen to the margin of error if Gallup had surveyed 100 people instead of the 1,051?

3. The Holiday Inn Resort Brand conducted the Kid Classified survey. 1,500 parents and children nationwide were interviewed via an online survey.

    The results of the survey state

    While many parents surveyed say they have some financial savings set aside specifically for vacation travel, more than half of parents in the survey (52%) noted that saving enough money was the biggest challenge to planning a family vacation, more so than coordination of family schedules (19%) or taking time off of work (12%).

    Source: http://www.lodgingmagazine.com/holiday-inn-resorts-catering-to-kids/

    a. Calculate the margin of error associated with the estimate of the proportion of all parents who would say that saving enough money is the biggest challenge to planning a family vacation.

    b. Write a sentence interpreting the margin of error.

    c. Comment on how the survey was conducted.

Lesson 23: Experiments and the Role of Random Assignment

Classwork

Exercises 1–4: Experiments

Two studies are described below. One is an observational study, while the other is an experiment.

Study A:

A new dog food, specially designed for older dogs, has been developed. A veterinarian wants to test this new food against another dog food currently on the market to see if it improves dogs' health. Thirty older dogs were randomly assigned to either the "new" food group or the "current" food group. After they were fed either the "new" or "current" food for six months, their improvement in health was rated.

Study B:

The administration at a large school wanted to determine if there was a difference in the mean number of text messages sent by ninth-grade students and by eleventh-grade students during a day. Students in a random sample of 30 ninth-grade students were asked how many text messages they sent per day. Students in another random sample of 30 eleventh-grade students were asked how many text messages they sent per day. The difference in the mean number of texts per day was determined.

1. Which study is the experiment? Explain. Discuss the answer with your partner.

2. In your own words, describe what a subject is in an experiment.

3. In your own words, describe what a response variable is in an experiment.

4. In your own words, describe what a treatment is in an experiment.

Exercises 5–9: Random Selection and Random Assignment

Take another look at the two studies described above. Study A (the dog food study) is an experiment, while study B (text messages) is an observational study. The term *random sample* implies that a sample was randomly selected from a population. The terms *random selection* and *random assignment* have very different meanings.

Random selection refers to randomly selecting a sample from a population. Random selection allows generalization to a population and is used in well-designed observational studies. Sometimes, but not always, the subjects in an experiment are randomly selected.

Random assignment refers to randomly assigning the subjects in an experiment to treatments. Random assignment allows for cause-and-effect conclusions and is used in well-designed experiments.

In study B, the data were collected from two random samples of students.

5. Can the results of the survey be generalized to all ninth-grade and all eleventh-grade students at the school? Why or why not? Discuss the answer with your partner.

6. Suppose there really is a difference in the mean number of texts sent by ninth-grade students and by eleventh-grade students. Can we say that the grade level of the students is the cause of the difference in the mean number of texts sent? Why or why not? Discuss the answer with your partner.

In study A, the dogs were randomly assigned to one of the two types of food.

7. Suppose the dogs that were fed the new food showed improved health. Can we say that the new food is the cause of the improvement in the dogs' health? Why or why not? Discuss the answer with your partner.

8. Can the results of the dog food study be generalized to all dogs? To all older dogs? Why or why not? Discuss the answer with your partner.

The table below summarizes the differences between the terms *random selection* and *random assignment*.

9. For each statement, put a check mark in the appropriate column(s), and explain your choices.

	Random Selection	Random Assignment
Used in Experiments		
Used in Observational Studies		
Allows Generalization to the Population		
Allows a Cause-and-Effect Conclusion		

Exercises 10–17

What is the purpose of random assignment in experiments? To answer this, consider the following investigation:

A researcher wants to determine if the yield of corn is different when the soil is treated with one of two different types of fertilizers, fertilizer A and fertilizer B. The researcher has 16 acres of land located beside a river that has several trees along its bank. There are also a few trees to the north of the 16 acres. The land has been divided into 16 one-acre plots. (See the diagram below.) These 16 plots are to be planted with the same type of corn but can be fertilized differently. At the end of the growing season, the corn yield will be measured for each plot, and the mean yields for the plots assigned to each fertilizer will be compared.

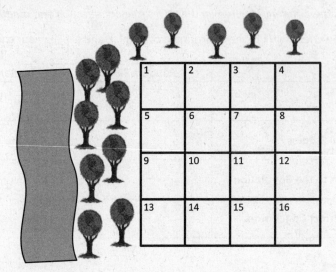

10. For the experiment, identify the following, and explain each answer:

   a.   Subjects (Hint: not always people or animals)

   b.   Treatments

   c.   Response variable

EUREKA
MATH™

Next, you need to assign the plots to one of the two treatments. To do this, follow the instructions given by your teacher.

11. Write A (for fertilizer A) or B (for fertilizer B) in each of the 16 squares in the diagram so that it corresponds to your random assignment of fertilizer to plots.

Let's investigate the results of the random assignment of the fertilizer types to the plots.

12. On the diagram above, draw a vertical line down the center of the 16 plots of land.

13. Count the number of plots on the left side of the vertical line that will receive fertilizer A. Count the number of plots on the right side of the vertical line that will receive fertilizer A.

                    Left _____   Right _____

14. On the diagram above, draw a horizontal line through the center of the 16 plots of land.

15. Count the number of plots above the horizontal line that will receive fertilizer A. Count the number of plots below the horizontal line that will receive fertilizer A.

                    Above _____   Below _____

In experiments, random assignment is used as a way of ensuring that the groups that receive each treatment are as much alike as possible with respect to other factors that might affect the response.

16. Explain what this means in the context of this experiment.

17. Suppose that, at the end of the experiment, the mean yield for one of the fertilizers is quite a bit higher than the mean yield for the other fertilizer. Explain why it would be reasonable to say that the type of fertilizer is the cause of the difference in yield and not the proximity to the river or to the northern trees.

Lesson Summary

- An *experiment* is an investigation designed to compare the effect of two or more treatments on a response variable.

- A *subject* is a participant in the experiment.

- The *response variable* is a variable that is not controlled by the experimenter and that is measured as part of the experiment.

- The *treatments* are the conditions to which subjects are randomly assigned by the experimenter.

- *Random selection* refers to randomly selecting a sample from a population.

  □ Random selection allows for generalization to a population.

- *Random assignment* refers to randomly assigning subjects to treatment groups.

  □ Random assignment allows for cause-and-effect conclusions.

  □ The purpose of random assignment in an experiment is to create similar groups of subjects for each of the treatments in the experiment.

Problem Set

For Problems 1 through 5, identify (i) the subjects, (ii) the treatments, and (iii) the response variable for each experiment.

1. A botanist was interested in determining the effects of watering (three days a week or daily) on the heat rating of jalapeño peppers. The botanist wanted to know which watering schedule would produce the highest heat rating in the peppers. He conducted an experiment, randomly assigning each watering schedule to half of 12 plots that had similar soil and full sun. The average final heat rating for the peppers grown in each plot was recorded at the end of the growing season.

2. A manufacturer advertises that its new plastic cake pan bakes cakes more evenly. A consumer group wants to carry out an experiment to see if the plastic cake pans do bake more evenly than standard metal cake pans. Twenty cake mixes (same brand and type) are randomly assigned to either the plastic pan or the metal pan. All of the cakes are baked in the same oven. The rating scale was then used to rate the evenness of each cake.

3. The city council of a large city is considering a new law that prohibits talking on a cell phone while driving. A consumer rights organization wants to know if talking on a cell phone while driving distracts a person's attention, causing that person to make errors while driving. An experiment is designed that uses a driving simulator to compare the two treatments: driving while talking on a cell phone and driving while not talking on a cell phone. The number of errors made while driving on an obstacle course will be recorded for each driver. Each person in a random sample of 200 licensed drivers in the city was asked to participate in the experiment. All of the drivers agreed to participate in the experiment. Half of the drivers were randomly assigned to drive an obstacle course while talking on a phone. The remaining half were assigned to drive the obstacle course while not talking on a phone.

4. Researchers studied 208 infants whose brains were temporarily deprived of oxygen as a result of complications at birth (*The New England Journal of Medicine*, October 13, 2005). An experiment was performed to determine if reducing body temperature for three days after birth improved their chances of surviving without brain damage. Infants were randomly assigned to usual care or whole-body cooling. The amount of brain damage was measured for each infant.

5. The head of the quality control department at a printing company would like to carry out an experiment to determine which of three different glues results in the greatest binding strength. Copies of a book were randomly assigned to one of the three different glues.

6. In Problem 3, suppose that drivers who talked on the phone while driving committed more errors on the obstacle course than drivers who did not talk on the phone while driving. Can we say that talking on the cell phone while driving is the cause of the increased errors on the obstacle course? Why or why not?

7. Can the results of the experiment in Problem 3 be generalized to all licensed drivers in the city? Why or why not?

8. In Problem 4, one of the treatment groups was to use usual care for the infants. Why was this treatment group included in the experiment?

9. In Problem 5, why were copies of only one book used in the experiment?

This page intentionally left blank

Lesson 24: Differences Due to Random Assignment Alone

Classwork

Exercises 1–17

Twenty adult drivers were asked the following question:

"What speed is the fastest that you have driven?"

The table below summarizes the fastest speeds driven in miles per hour (mph).

70	60	70	95	50	60	80	75	55	90
110	65	65	65	55	70	75	70	65	40

1. What is the mean fastest speed driven?

2. What is the range of fastest speed driven?

3. Imagine that the fastest speeds were randomly divided into two groups. How would the means and ranges compare to one another? To the means and ranges of the whole group? Explain your thinking.

Let's investigate what happens when the fastest speeds driven are randomly divided into two equal-sized groups.

4. Following the instructions from your teacher, randomly divide the 20 values in the table above into two groups of 10 values each.

											Mean
Group 1											
Group 2											

5. Do you expect the means of these two groups to be equal? Why or why not?

6. Compute the means of these two groups. Write the means in the chart above.

7. How do these two means compare to each other?

8. How do these two means compare to the mean fastest speed driven for the entire group (Exercise 1)?

9. Use the instructions provided for Exercise 4 to repeat the random division process two more times. Compute the mean of each group for each of the random divisions into two groups. Record your results in the table below.

											Mean
Group 3											
Group 4											
Group 5											
Group 6											

10. Plot the means of all six groups on a class dot plot.

11. Based on the class dot plot, what can you say about the possible values of the group means?

12. What is the smallest possible value for a group mean? Largest possible value?

13. What is the largest possible range for the distribution of group means?

14. How does the largest possible range in the group means compare to the range of the original data set (Exercise 2)? Why is this so?

15. What is the shape of the distribution of group means?

16. Will your answer to the above question always be true? Explain.

17. When a single set of values is randomly divided into two equal groups, explain how the means of these two groups may be very different from each other and may be very different from the mean of the single set of values.

Lesson Summary

When a single set of values is randomly divided into two groups:

- The two group means will tend to differ just by chance.
- The distribution of random groups' means will be centered at the single set's mean.
- The range of the distribution of the random groups' means will be smaller than the range of the data set.
- The shape of the distribution of the random groups' means will be symmetrical.

Problem Set

In one high school, there are eight math classes during second period. The number of students in each second-period math class is recorded below.

32   27   26   23   25   22   30   19

This data set is randomly divided into two equal-sized groups, and the group means are computed.

1. Will the two group means be the same? Why or why not?

The random division into two groups process is repeated many times to create a distribution of group mean class size.

2. What is the center of the distribution of group mean class size?

3. What is the largest possible range of the distribution of group mean class size?

4. What possible values for the mean class size are more likely to happen than others? Explain why you chose these values.

There are 3 different sets of numbers: Set A, Set B, and Set C. Each set contains 10 numbers. In two of the sets, the 10 numbers were randomly divided into two groups of 5 numbers each, and the mean for each group was calculated. These two means are plotted on a dot plot. This procedure was repeated many times, and the dot plots of the group means are shown below.

The third set did not use the above procedure to compute the means.

For each set, the smallest possible group mean and the largest possible group mean were calculated, and these two means are shown in the dot plots below.

Use the dot plots below to answer Problems 5–8.

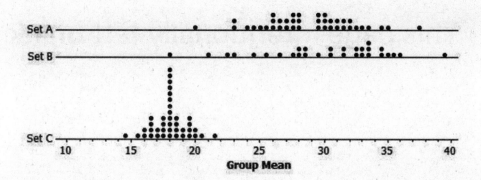

5. Which set is *not* one of the two sets that were randomly divided into two groups of 5 numbers? Explain.

6. Estimate the mean of the original values in Set A. Show your work.

7. Estimate the range of the group means shown in the dot plot for Set C. Show your work.

8. Is the range of the original values in Set C smaller or larger than your answer in Problem 7? Explain.

This page intentionally left blank

Appendix A

65	40	55	90
70	75	80	75
55	70	50	60
65	65	95	70
110	65	60	70

This page intentionally left blank

Lesson 25: Ruling Out Chance

Classwork

Opening Exercise

a. Explain in writing what you learned about randomly divided groups from the last lesson. Share your thoughts with a neighbor.

b. How could simulation be used to understand the typical differences between randomized groups?

Exercises 1–3: Random Assignments and Computing the Difference of the Group Means

Imagine that 10 tomatoes of varying shapes and sizes have been placed in front of you. These 10 tomatoes (all of the same variety) have been part of a nutrient experiment where the application of the nutrient is expected to yield larger tomatoes that weigh more. All 10 tomatoes have been grown under similar conditions regarding soil, water, and sunlight, but 5 of the tomatoes received the additional nutrient supplement. Using the weight data of these 10 tomatoes, you wish to examine the claim that the nutrient yields larger tomatoes on average.

1. Why would it be important in this experiment for the 10 tomatoes to all be of the same variety and grown under the same conditions (except for the treatment applied to 5 of the tomatoes)?

Here are the 10 tomatoes with their weights shown. They have been ordered from largest to smallest based on weight.

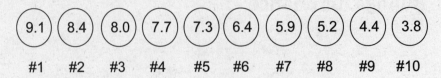

9.1 8.4 8.0 7.7 7.3 6.4 5.9 5.2 4.4 3.8

#1   #2   #3   #4   #5   #6   #7   #8   #9   #10

For now, do not be concerned about which tomatoes received the additional nutrients. The object here is to randomly assign the tomatoes to two groups.

Imagine that someone assisting you uses a random number generator or some other impartial selection device and randomly selects tomatoes 1, 4, 5, 7, and 10 to be in Group A. By default, tomatoes 2, 3, 6, 8, and 9 will be in Group B. The result is illustrated below.

Group A	Group B
9.1	8.4
7.7	8.0
7.3	6.4
5.9	5.2
3.8	4.4

2. Confirm that the mean for Group A is 6.76 ounces, and calculate the mean for Group B.

3. Calculate the difference between the mean of Group A and the mean of Group B (that is, calculate $\bar{x}_A - \bar{x}_B$).

EUREKA
MATH™

Exercises 4–6: Interpreting the Value of a Difference

The statistic of interest that you care about is the difference between the mean of the 5 tomatoes in Group A and the mean of the 5 tomatoes in Group B. For now, call that difference *Diff*: $\text{Diff} = \bar{x}_A - \bar{x}_B$

4. Explain what a Diff value of 1.64 ounces would mean in terms of which group has the larger mean weight and the number of ounces by which that group's mean weight exceeds the other group's mean weight.

5. Explain what a Diff value of -0.4 ounces would mean in terms of which group has the larger mean weight and the number of ounces by which that group's mean weight exceeds the other group's mean weight.

6. Explain what a Diff value of 0 ounces would mean regarding the difference between the mean weight of the 5 tomatoes in Group A and the mean weight of the 5 tomatoes in Group B.

EUREKA
MATH™

Lesson 25:   Ruling Out Chance

S.205

©2015 Great Minds. eureka-math.org
ALG II-M4-SE-B2-1.3.0-08.2015

Exercises 7–8: Additional Random Assignments

7. Below is a second random assignment of the 10 tomatoes to two groups. Calculate the mean of each group, and then calculate the value of Diff for this second case. Also, interpret the Diff value in context using your responses to the previous questions as a guide.

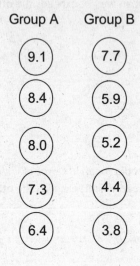

Group A   Group B

9.1   7.7

8.4   5.9

8.0   5.2

7.3   4.4

6.4   3.8

8. Here is a third random assignment of the 10 tomatoes. Calculate the mean of each group, and then calculate the value of Diff for this case. Interpret the Diff value in context using your responses to the previous questions as a guide.

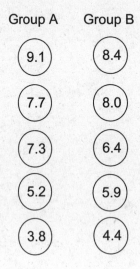

Group A   Group B

9.1   8.4

7.7   8.0

7.3   6.4

5.2   5.9

3.8   4.4

EUREKA
MATH™

©2015 Great Minds. eureka-math.org
ALG II-M4-SE-B2-1.3.0-08.2015

Lesson Summary

In this lesson, when the single group of observations was randomly divided into two groups, the means of these two groups differed by chance. These differences have a context based on the purpose of the experiment and the units of the original observations.

The differences varied. In some cases, the difference in the means of these two groups was very small (or 0), but in other cases, this difference was larger. However, in order to determine which differences were typical and ordinary versus unusual and rare, a sense of the center, spread, and shape of the distribution of possible differences is needed. In the following lessons, you will develop this distribution by executing repeated random assignments similar to the ones you saw in this lesson.

Problem Set

Six ping-pong balls are labeled as follows: $0, 3, 6, 9, 12, 18$. Three ping-pong balls will be randomly assigned to Group A; the rest will be assigned to Group B. Recall that $\text{Diff} = \bar{x}_A - \bar{x}_B$

In the Exit Ticket problem, 4 of the 20 possible randomizations have been addressed; those possibilities are shown in the first four rows of the table below.

1. Develop the remaining 16 possible random assignments to two groups, and calculate the Diff value for each.

    (Note: Avoid redundant cases; selecting 0, 3, and 6 for Group A is *not* a distinct random assignment from selecting 6, 0, and 3; do not record both.)

Group A Selection			$\bar{x}_A$	$\bar{x}_B$	Diff	
3	6	12	7	9	-2	Question #1
3	12	18	11	5	6	Question #2
9	12	18	13	3	10	Question #3
0	3	6	3	13	-10	Question #4

©2015 Great Minds. eureka-math.org
ALG II-M4-SE-B2-1.3.0-08.2015

2. Create a dot plot that shows the 20 Diff values obtained from the 20 possible randomizations. By visual inspection, what is the mean and median value of the distribution?

3. Based on your dot plot, what is the probability of obtaining a Diff value of 8 or higher?

4. Would a Diff value of 8 or higher be considered a difference that is likely to happen or one that is unlikely to happen? Explain.

5. Based on your dot plot, what is the probability of obtaining a Diff value of −2 or smaller?

6. Would a Diff value of −2 or smaller be considered a difference that is likely to happen or one that is unlikely to happen? Explain.

©2015 Great Minds. eureka-math.org
ALG II-M4-SE-B2-1.3.0-08.2015

This page intentionally left blank

Lesson 26: Ruling Out Chance

Classwork

Opening Exercise

Previously, you considered the random assignment of 10 tomatoes into two distinct groups of 5 tomatoes, Group A and Group B. With each random assignment, you calculated $\text{Diff} = \bar{x}_A - \bar{x}_B$, the difference between the mean weight of the 5 tomatoes in Group A and the mean weight of the 5 tomatoes in Group B.

    a.   Summarize in writing what you learned in the last lesson. Share your thoughts with a neighbor.

    b.   Recall that 5 of these 10 tomatoes are from plants that received a nutrient treatment in the hope of growing bigger tomatoes. But what if the treatment was *not* effective? What difference would you expect to find between the group means?

Exercises 1–2: The Distribution of Diff and Why 0 Is Important

In the previous lesson, 3 instances of the tomato randomization were considered. Imagine that the random assignment was conducted an additional 247 times, and 250 Diff values were computed from these 250 random assignments. The results are shown graphically below in a dot plot where each dot represents the Diff value that results from a random assignment.

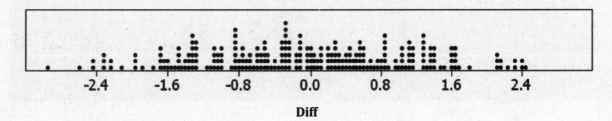

Diff

This dot plot will serve as your *randomization distribution* for the Diff statistic in this tomato randomization example. The dots are placed at increments of 0.04 ounces.

1. Given the distribution picture above, what is the *approximate* value of the median and mean of the distribution? Specifically, do you think this distribution is centered near a value that implies "No Difference" between Group A and Group B?

2. Given the distribution pictured above and based on the simulation results, determine the approximate probability of obtaining a Diff value in the cases described in (a), (b), and (c).

   a. Of 1.64 ounces or more

   b. Of −0.80 ounces or less

   c. Within 0.80 ounces of 0 ounces

   d. How do you think these probabilities could be useful to people who are designing experiments?

Exercises 3–5: Statistically Significant Diff Values

In the context of a randomization distribution that is based upon the assumption that there is no real difference between the groups, consider a Diff value of X to be statistically significant if there is a low probability of obtaining a result that is as extreme as or more extreme than X.

3. Using that definition and your work above, would you consider any of the Diff values below to be statistically significant? Explain.

   a. 1.64 ounces

   b. −0.80 ounces

   c. Values within 0.80 ounces of 0 ounces

4. In the previous lessons, you obtained Diff values of 0.28 ounces, 2.44 ounces, and 0 ounces for 3 different tomato randomizations. Would you consider any of those values to be statistically significant for this distribution? Explain.

5. Recalling that Diff is the mean weight of the 5 Group A tomatoes minus the mean weight of the 5 Group B tomatoes, how would you explain the meaning of a Diff value of 1.64 ounces in this case?

Exercises 6–8: The Implication of Statistically Significant Diff Values

Keep in mind that for reasons mentioned earlier, the randomization distribution above is demonstrating what is likely to happen *by chance alone* if the treatment was *not* effective. As stated in the previous lesson, you can use this randomization distribution to assess whether or not the *actual* difference in means *obtained from your experiment* (the difference between the mean weight of the 5 actual control group tomatoes and the mean weight of the 5 actual treatment group tomatoes) is consistent with usual chance behavior. The logic is as follows:

- If the observed difference is "extreme" and not typical of chance behavior, it may be considered statistically significant and possibly not the result of chance behavior.

- If the difference is not the result of chance behavior, then maybe the difference did not just happen by chance alone.

- If the difference did not just happen by chance alone, maybe the difference you observed is caused by the treatment in question, which, in this case, is the nutrient. In the context of our example, a statistically significant Diff value provides evidence that the nutrient treatment did in fact yield heavier tomatoes on average.

6. For reasons that will be explained in the next lesson, for your tomato example, Diff values that are *positive* and statistically significant will be considered as good evidence that your nutrient treatment did in fact yield heavier tomatoes on average. Again, using the randomization distribution shown earlier in the lesson, which (if any) of the following Diff values would you consider to be statistically significant and lead you to think that the nutrient treatment did in fact yield heavier tomatoes on average? Explain for each case.

$$\text{Diff} = 0.4, \text{Diff} = 0.8, \text{Diff} = 1.2, \text{Diff} = 1.6, \text{Diff} = 2.0, \text{Diff} = 2.4$$

©2015 Great Minds. eureka-math.org
ALG II-M4-SE-B2-1.3.0-08.2015

7. In the first random assignment in the previous lesson, you obtained a Diff value of 0.28 ounces. Earlier in this lesson, you were asked to consider if this might be a statistically significant value. Given the distribution shown in this lesson, if you had obtained a Diff value of 0.28 ounces *in your experiment* and the 5 Group A tomatoes had been the "treatment" tomatoes that received the nutrient, would you say that the Diff value was extreme enough to support a conclusion that the nutrient treatment yielded heavier tomatoes on average? Or do you think such a Diff value may just occur by chance when the treatment is ineffective? Explain.

8. In the second random assignment in the previous lesson, you obtained a Diff value of 2.44 ounces. Earlier in this lesson, you were asked to consider if this might be a statistically significant value. Given the distribution shown in this lesson, if you had obtained a Diff value of 2.44 ounces *in your experiment* and the 5 Group A tomatoes had been the "treatment" tomatoes that received the nutrient, would you say that the Diff value was extreme enough to support a conclusion that the nutrient treatment yielded heavier tomatoes on average? Or do you think such a Diff value may just occur by chance when the treatment is ineffective? Explain.

©2015 Great Minds. eureka-math.org
ALG II-M4-SE-B2-1.3.0-08.2015

Lesson Summary

In the previous lesson, the concept of randomly separating 10 tomatoes into 2 groups and comparing the means of each group was introduced. The randomization distribution of the difference in means that is created from multiple occurrences of these random assignments demonstrates what is likely to happen *by chance alone* if the nutrient treatment is *not* effective. When the results of your tomato growth experiment are compared to that distribution, you can then determine if the tomato growth experiment's results were typical of chance behavior.

If the results appear typical of chance behavior and near the center of the distribution (that is, not relatively very far from a Diff of 0), then there is little evidence that the treatment was effective. However, if it appears that the experiment's results are not typical of chance behavior, then maybe the difference you are observing didn't just happen by chance alone. It may indicate a statistically significant difference between the treatment group and the control group, and the source of that difference might be (in this case) the nutrient treatment.

Problem Set

In each of the 3 cases below, calculate the Diff value as directed, and write a sentence explaining what the Diff value means in context. Write the sentence for a general audience.

1. Group A: 8 dieters lost an average of 8 pounds, so $\bar{x}_A = -8$.

    Group B: 8 nondieters lost an average of 2 pounds over the same time period, so $\bar{x}_B = -2$.

    Calculate and interpret Diff $= \bar{x}_A - \bar{x}_B$.

2. Group A: 11 students were on average 0.4 seconds faster in their 100-meter run times after following a new training regimen.

    Group B: 11 students were on average 0.2 seconds slower in their 100-meter run times after not following any new training regimens.

    Calculate and interpret Diff $= \bar{x}_A - \bar{x}_B$.

3. Group A: 20 squash that have been grown in an irrigated field have an average weight of 1.3 pounds.

    Group B: 20 squash that have been grown in a nonirrigated field have an average weight of 1.2 pounds.

    Calculate and interpret Diff $= \bar{x}_A - \bar{x}_B$.

4. Using the randomization distribution shown below, what is the probability of obtaining a Diff value of -0.6 or less?

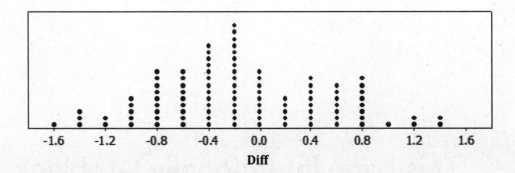

5. Would a Diff value of -0.6 or less be considered a statistically significant difference? Why or why not?

6. Using the randomization distribution shown in Problem 4, what is the probability of obtaining a Diff value of -1.2 or less?

7. Would a Diff value of -1.2 or less be considered a statistically significant difference? Why or why not?

This page intentionally left blank

Lesson 27: Ruling Out Chance

Classwork

Exercises 1–4: Carrying Out a Randomization Test

The following are the general steps for carrying out a randomization test to analyze the results of an experiment. The steps are also presented in the context of the tomato example of the previous lessons.

<u>*Step 1—Develop competing claims: no difference versus difference.*</u>

One claim corresponds to no difference between the two groups in the experiment. This claim is called the *null hypothesis*.

- For the tomato example, the null hypothesis is that the nutrient treatment is not effective in increasing tomato weight. This is equivalent to saying that the average weight of treated tomatoes may be the same as the average weight of nontreated (control) tomatoes.

The competing claim corresponds to a difference between the two groups. This claim could take the form of a *different from*, *greater than*, or *less than* statement. This claim is called the *alternative hypothesis*.

- For the tomato example, the alternative hypothesis is that the nutrient treatment is effective in increasing tomato weight. This is equivalent to saying that the average weight of treated tomatoes *is greater than* the average weight of nontreated (control) tomatoes.

1. Previously, the statistic of interest that you used was the difference between the mean weight of the 5 tomatoes in Group A and the mean weight of the 5 tomatoes in Group B. That difference was called *Diff*. $\text{Diff} = \bar{x}_A - \bar{x}_B$ If the treatment tomatoes are represented by Group A and the control tomatoes are represented by Group B, what type of statistically significant values of Diff would support the claim that the average weight of treated tomatoes *is greater than* the average weight of nontreated (control) tomatoes: negative values of Diff, positive values of Diff, or both? Explain.

Step 2—*Take measurements from each group, and calculate the value of the* Diff *statistic from the experiment.*

For the tomato example, first, measure the weights of the 5 tomatoes from the treatment group (Group A); next, measure the weights of the 5 tomatoes from the control group (Group B); finally, compute $\text{Diff} = \bar{x}_A - \bar{x}_B$, which will serve as the result from your experiment.

2. Assume that the following represents the two groups of tomatoes from the *actual* experiment. Calculate the value of $\text{Diff} = \bar{x}_A - \bar{x}_B$. This will serve as the result from your experiment.

    These are the same 10 tomatoes used in previous lessons; the identification of which tomatoes are treatment versus control is now revealed.

TREATMENT Group A	CONTROL Group B
9.1	7.7
8.4	6.4
8.0	5.2
7.3	4.4
5.9	3.8

Again, these tomatoes represent the actual result from your experiment. You will now create the randomization distribution by making repeated random assignments of these 10 tomatoes into 2 groups and recording the observed difference in means for each random assignment. This develops a randomization distribution of the many possible difference values that could occur under the assumption that there is no difference between the mean weights of tomatoes that receive the treatment and tomatoes that don't receive the treatment.

EUREKA
MATH™

Step 3—*Randomly assign the observations to two groups, and calculate the difference between the group means. Repeat this several times, recording each difference. This will create the randomization distribution for the* Diff *statistic.*

Examples of this technique were presented in a previous lesson. For the tomato example, the randomization distribution has already been presented in a previous lesson and is shown again here. The dots are placed at increments of 0.04 ounce.

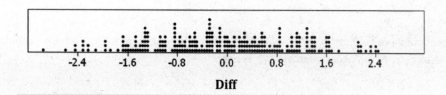

Diff

Step 4—*With reference to the randomization distribution (Step 3) and the inequality in your alternative hypothesis (Step 1), compute the probability of getting a* Diff *value as extreme as or more extreme than the* Diff *value you obtained in your experiment (Step 2).*

For the tomato example, since the treatment group is Group A, the Diff value of $\bar{x}_A - \bar{x}_B$ is $\bar{x}_{\text{Treatment}} - \bar{x}_{\text{Control}}$. Since the alternative claim is supported by $\bar{x}_{\text{Treatment}} > \bar{x}_{\text{Control}}$, you are seeking statistically significant Diff values that are positive since if $\bar{x}_{\text{Treatment}} > \bar{x}_{\text{Control}}$, then $\bar{x}_{\text{Treatment}} - \bar{x}_{\text{Control}} > 0$.

Statistically significant values of Diff that are negative, in this case, would imply that the treatment made the tomatoes smaller on average.

3.   Using your calculation from Exercise 2, determine the probability of getting a Diff value as extreme as or more extreme than the Diff value you obtained for this experiment (in Step 2).

Step 5—*Make a conclusion in context based on the probability calculation (Step 4).*

If there is a *small probability* of obtaining a Diff value as extreme as or more extreme than the Diff value you obtained in your experiment, then *the* Diff *value from the experiment is unusual* and not typical of chance behavior. Your experiment's results probably did not happen by chance, and the results probably occurred because of a statistically significant difference in the two groups.

-   In the tomato experiment, if you think there is a statistically significant difference in the two groups, *you have evidence that the treatment may in fact be yielding heavier tomatoes on average.*

If there is *not a small probability* of obtaining a Diff value as extreme as or more extreme than the Diff value you obtained in your experiment, then *your* Diff *value from the experiment is NOT considered unusual* and could be typical of chance behavior. The experiment's results may have just happened by chance and not because of a statistically significant difference in the two groups.

- In the tomato experiment, if you don't think that there is a statistically significant difference in the two groups, then you do *not* have evidence that the treatment results in larger tomatoes on average.

In some cases, a specific cutoff value called a *significance level* might be employed to assist in determining how small this probability must be in order to consider results statistically significant.

4. Based on your probability calculation in Exercise 3, do the data from the tomato experiment support the claim that the treatment yields heavier tomatoes on average? Explain.

Exercises 5–10: Developing the Randomization Distribution

Although you are familiar with how a randomization distribution is created in the tomato example, the randomization distribution was provided for you. In this exercise, you will develop two randomization distributions based on the same group of 10 tomatoes. One distribution will be developed by hand and will contain the results of at least 250 random assignments. The second distribution will be developed using technology and will contain the results of at least 250 random assignments. Once the two distributions have been developed, you will be asked to compare the distributions.

Manually Generated

Your instructor will provide you with specific guidance regarding how many random assignments you need to carry out. Ultimately, your class should generate at least 250 random assignments, compute the Diff value for each, and record these 250 or more Diff values on a class or an individual dot plot.

5. To begin, write the 10 tomato weights on 10 equally sized slips of paper, one weight on each slip. Place the slips in a container, and shake the container well. Remove 5 slips, and assign those 5 tomatoes as Group A. The remaining tomatoes will serve as Group B.

6. Calculate the mean weight for Group A and the mean weight for Group B. Then, calculate Diff $= \bar{x}_A - \bar{x}_B$ for this random assignment.

7. Record your Diff value, and add this value to the dot plot. Repeat as needed per your instructor's request until a manually generated randomization distribution of at least 250 differences has been achieved.

    (Note: This distribution will most likely be slightly different from the tomato randomization distribution given earlier in this lesson.)

Computer Generated

At this stage, you will be encouraged to use a Web-based randomization testing applet/calculator to perform the steps above. The applet is located at http://www.rossmanchance.com/applets/AnovaShuffle.htm. To supplement the instructions below, a screenshot of the applet appears as the final page of this lesson.

Upon reaching the applet, do the following:

- Press the Clear button to clear the data under Sample Data.
- Enter the tomato data exactly as shown below. When finished, press the Use Data button.

Group	Ounces
Treatment	9.1
Treatment	8.4
Treatment	8
Control	7.7
Treatment	7.3
Control	6.4
Treatment	5.9
Control	5.2
Control	4.4
Control	3.8

Once the data are entered, notice that dot plots of the two groups appear. Also, the statistic window below the data now says "difference in means," and an Observed Diff value of 2.24 is computed for the experiment's data (just as you computed in Exercise 2).

By design, the applet will determine the difference of means based on the first group name it encounters in the data set—specifically, it will use the first group name it encounters as the first value in the difference of means calculation. In other words, to compute the difference in means as $\bar{x}_{\text{Treatment}} - \bar{x}_{\text{Control}}$, a Treatment observation needs to appear prior to any control observations in the data set as entered.

- Select the check box next to Show Shuffle Options, and a dot plot template will appear.

- Enter 250 in the box next to Number of Shuffles, and press the Shuffle Responses button. A randomization distribution based on 250 randomizations (in the form of a histogram) is created.

  This distribution will most likely be *slightly* different from both the tomato randomization distribution that appeared earlier in this lesson and the randomization distribution that was manually generated in Exercise 7.

8. Write a few comments comparing the manually generated distribution and the computer-generated distribution. Specifically, did they appear to have roughly the same shape, center, and spread?

The applet also allows you to compute probabilities. For this case:

- Under Count Samples, select Greater Than. Then, in the box next to Greater Than, enter 2.2399.

  Since the applet computes the count value as *strictly* greater than and not greater than or equal to, in order to obtain the probability of obtaining a value as extreme as or more extreme than the Observed Diff value of 2.24, you will need to enter a value just slightly below 2.24 to ensure that Diff observations of 2.24 are included in the count.

- Select the Count button. The probability of obtaining a Diff value of 2.24 or more in this distribution will be computed for you.

  The applet displays the randomization distribution in the form of a histogram, and it shades in red *all* histogram classes that contain *any* difference values that meet your Count Samples criteria. Due to the grouping and binning of the classes, some of the red shaded classes (bars) may also contain difference values that do not fit your Count Samples criteria. Just keep in mind that the Count value stated in red below the histogram will be exact; the red shading in the histogram may be approximate.

9. How did the probability of obtaining a Diff value of 2.24 or more using your computer-generated distribution compare with the probability of obtaining a Diff value of 2.24 or more using your manually generated distribution?

10. Would you come to the same conclusion regarding the experiment using either the computer-generated or manually generated distribution? Explain. Is this the same conclusion you came to using the distribution shown earlier in this lesson back in Step 3?

Lesson Summary

The following are the general steps for carrying out a randomization test to analyze the results of an experiment:

Step 1—Develop competing claims: no difference versus difference.

Develop the null hypothesis: This claim is that there is no difference between the two groups in the experiment.

Develop the alternative hypothesis: The competing claim is that there *is* a difference between the two groups. This difference could take the form of a *different from*, *greater than*, or *less than* statement depending on the purpose of the experiment and the claim being assessed.

Step 2—Take measurements from each group, and calculate the value of the Diff *statistic from the experiment.*

This is the observed Diff value from the experiment.

Step 3—Randomly assign the observations to two groups, and calculate the difference between the group means. Repeat this several times, recording each difference.

This will create the *randomization distribution* for the Diff statistic under the assumption that there is no statistically significant difference between the two groups.

Step 4—With reference to the randomization distribution (from Step 3) and the inequality in your alternative hypothesis (from Step 1), compute the probability of getting a Diff *value as extreme as or more extreme than the* Diff *value you obtained in your experiment (from Step 2).*

Step 5—Make a conclusion in context based on the probability calculation (from Step 4).

Small probability: If the Diff value from the experiment is unusual and not typical of chance behavior, your experiment's results probably did not happen by chance. The results probably occurred because of a statistically significant difference in the two groups.

Not a small probability: If the Diff value from the experiment is *not* considered unusual and could be typical of chance behavior, your experiment's results may have just happened by chance and *not* because of a statistically significant difference in the two groups.

Note: The use of technology is strongly encouraged to assist in Steps 2–4.

©2015 Great Minds. eureka-math.org
ALG II-M4-SE-B2-1.3.0-08.2015

Problem Set

1. Using the 20 observations that appear in the table below for the changes in pain scores of 20 individuals, use the Anova Shuffle applet to develop a randomization distribution of the value $\text{Diff} = \bar{x}_A - \bar{x}_B$ based on 100 random assignments of these 20 observations into two groups of 10. Enter the data exactly as shown below. Describe similarities and differences between this new randomization distribution and the distribution shown.

Group	ChangeinScore
A	0
A	0
A	-1
A	-1
A	-2
A	-2
A	-3
A	-3
A	-3
A	-4
B	0
B	0
B	0
B	0
B	0
B	0
B	-1
B	-1
B	-1
B	-2

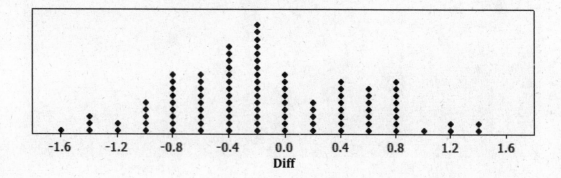

2. In a previous lesson, the burn times of 6 candles were presented. It is believed that candles from Group A will burn longer on average than candles from Group B. The data from the experiment (now shown with group identifiers) are provided below.

Group	Burntime
A	18
A	12
B	9
A	6
B	3
B	0

Perform a randomization test of this claim. Carry out all 5 steps, and use the Anova Shuffle applet to perform Steps 2–4. Enter the data exactly as presented above, and in Step 3, develop the randomization distribution based on 200 random assignments.

EUREKA
MATH™

©2015 Great Minds. eureka-math.org
ALG II-M4-SE-B2-1.3.0-08.2015

Lesson 28: Drawing a Conclusion from an Experiment

Classwork

In this lesson, you will be conducting all phases of an experiment: collecting data, creating a randomization distribution based on these data, and determining if there is a significant difference in treatment effects. In the next lesson, you will develop a report of your findings.

The following experiments are in homage to George E. P. Box, a famous statistician who worked extensively in the areas of quality control, design of experiments, and other topics. He earned the honor of Fellow of the Royal Society during his career and is a former president of the American Statistical Association. Several resources are available regarding his work and life including the book *Statistics for Experimenters: Design, Innovation, and Discovery* by Box, Hunter, and Hunter.

The experiments will investigate whether modifications in certain dimensions of a paper helicopter will affect its flight time.

Exercise: Build the Helicopters

In preparation for your data collection, you will need to construct 20 paper helicopters following the blueprint given at the end of this lesson. For consistency, use the same type of paper for each helicopter. For greater stability, you may want to use a piece of tape to secure the two folded body panels to the body of the helicopter. By design, there will be some overlap from this folding in some helicopters.

You will carry out an experiment to investigate the effect of wing length on flight time.

a.   Construct 20 helicopters with wing length of 4 inches and body length of 3 inches. Label 10 each of these helicopters with the word *long*.

b.   Take the other 10 helicopters, and cut 1 inch off each of the wings so that you have 10 helicopters with 3-inch wings. Label each of these helicopters with the word *short*.

c.   How do you think wing length will affect flight time? Explain your answer.

©2015 Great Minds. eureka-math.org
ALG II-M4-SE-B2-1.3.0-08.2015

Exploratory Challenge 1: Data Collection

Once you have built the 20 helicopters, each of them will be flown by dropping the helicopter from a fixed distance above the ground (preferably 12 feet or higher—record this height for use when presenting your findings later). For consistency, drop all helicopters from the same height each time, and try to perform this exercise in a space where possible confounding factors such as wind gusts and drafts from heating and air conditioning are eliminated.

a. Place the 20 helicopters in a bag, shake the bag, and randomly pull out one helicopter. Drop the helicopter from the starting height and, using a stopwatch, record the amount of time it takes until the helicopter reaches the ground. Write down this flight time in the appropriate column in the table below. Repeat for the remaining 19 helicopters.

Some helicopters might fly more smoothly than others; you may want to record relevant comments in your report.

Flight Time (seconds)	
Long Wings (Group A)	Short Wings (Group B)

b. Why might it be important to randomize (impartially select) the order in which the helicopters were dropped?

(This is *different from* the randomization you perform later when you are allocating observations to groups to develop the randomization distribution.)

EUREKA
MATH™

©2015 Great Minds. eureka-math.org
ALG II-M4-SE-B2-1.3.0-08.2015

Exploratory Challenge 2: Developing Claims and Using Technology

With the data in hand, you will now perform your analysis regarding the effect of wing length.

Experiment: Wing Length

In this experiment, you will examine whether wing length makes a difference in flight time. You will compare the helicopters with long wings (wing length of 4 inches, Group A) to the helicopters with short wings (wing length of 3 inches, Group B). Since you are dropping the helicopters from the same height in the same location, using the same type of paper, the only difference in the two groups will be the different wing lengths.

Questions: Does a 1-inch addition in wing length appear to result in a change in average flight time? If so, do helicopters with longer wing length or shorter wing length tend to have longer flight times on average?

Carry out a complete randomization test to answer these questions. Show all 5 steps, and use the Anova Shuffle applet described in the previous lessons to assist both in creating the distribution and with your computations. Be sure to write a final conclusion that clearly answers the questions in context.

Lesson Summary

In previous lessons, you learned how to carry out a randomization test to decide if there was a statistically significant difference between two groups in an experiment. Throughout these previous lessons, certain aspects of proper experimental design were discussed. In this lesson, you were able to carry out a complete experiment and collect your own data. When an experiment is developed, you must be careful to minimize confounding effects that may compromise or invalidate findings. When possible, the treatment groups should be created so that the only distinction between the groups in the experiment is the treatment imposed.

Problem Set

One other variable that can be adjusted in the paper helicopters is body width. See the blueprints for details.

1. Construct 10 helicopters using the blueprint from the lesson. Label each helicopter with the word *narrow*.

2. Develop a blueprint for a helicopter that is identical to the blueprint used in class except for the fact that the body width will now be 1.75 inches.

3. Use the blueprint to construct 10 of these new helicopters, and label each of these helicopters with the word *wide*.

4. Place the 20 helicopters in a bag, shake the bag, and randomly pull out one helicopter. Drop the helicopter from the starting height and, using a stopwatch, record the amount of time it takes until the helicopter reaches the ground. Write down this flight time in the appropriate column in the table below. Repeat for the remaining 19 helicopters.

Flight Time (seconds)	
Narrow Body (Group C)	**Wide Body (Group D)**

5. Questions: Does a 0.5-inch addition in body width appear to result in a change in average flight time? If so, do helicopters with wider body width (Group D) or narrower body width (Group C) tend to have longer flight times on average? Carry out a complete randomization test to answer these questions. Show all 5 steps, and use the Anova Shuffle applet described in previous lessons to assist both in creating the distribution and with your computations. Be sure to write a final conclusion that clearly answers the questions in context.

Appendix: Blueprint

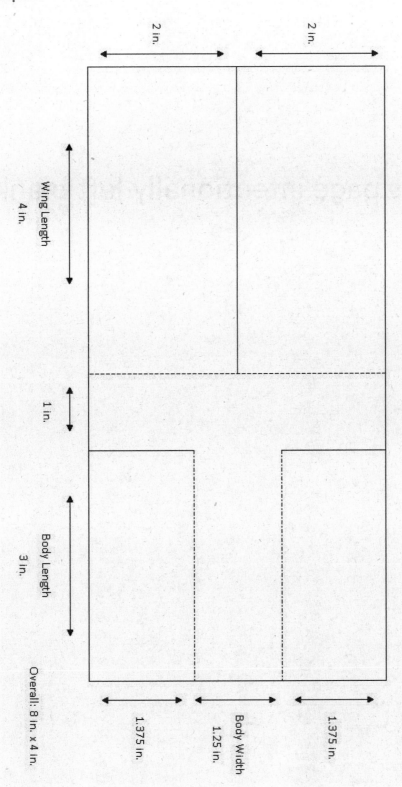

This page intentionally left blank

Lesson 29: Drawing a Conclusion from an Experiment

Classwork

In this lesson, you will develop a comprehensive poster summarizing your experiments.

Characteristics of a Good Poster

Your instructor will provide you with specific instructions and a rubric for assessing your poster (taken from "Poster Judging Rubric" at the "Poster Competition and Project Competition" page of the American Statistical Association, www.amstat.org/education/posterprojects/pdfs/PosterJudgingRubric.pdf).

Generally speaking, the presentation of a statistical analysis and/or an experiment should clearly state the question or purpose. The presentation should lead to the conclusion on a path that is easy to follow. The results of the study should be immediately obvious to the viewer. Any graphs included should be relevant to the question of interest and appropriate for the type of data collected.

Exploratory Challenge: Explaining the Experiment and Results

Your classwork will involve developing your poster. Your instructor will provide guidance as to groups, amount of time to spend, the rubric to be used for evaluation, etc. Your poster should address the results of both Experiments 1 and 2 regarding the effects of both body width and wing length.

In addition to the general concerns of colors, fonts to use, etc., in preparation for creating your poster, consider (and answer) these classwork questions.

- What was the objective of the experiment?

- How did you collect your data?

- What summary values and graphs should you present?

- How will you develop and present a summary of the experiment in a way that it is easy to follow and effortlessly leads the viewer to the conclusion?

- How will you explain *statistical significance*?

This page intentionally left blank

Lesson 30: Evaluating Reports Based on Data from an Experiment

Classwork

Exercises 1–7

Pericarditis is an inflammation (irritation and swelling) of the pericardium, the thin sac that surrounds the heart. When extra fluid builds up between the two layers of the pericardium, the heart's actions are restricted. An experiment reported in the article "A Randomized Trial of Colchicine for Acute Pericarditis" in *The New England Journal of Medicine* (October 2013) tested the effects of the drug colchicine on acute pericarditis.

Read the abstract of the article, and answer the following questions:

Website: www.nejm.org/doi/full/10.1056/NEJMoa1208536

1. How many treatment groups are there?

2. What treatments are being compared?

3. Is there a placebo group? Explain.

4. How many subjects are in each treatment group?

5. Do you think that the number of subjects in each treatment is enough? Explain.

6. What method was used to assign the subjects to the treatment groups? Explain why this is important.

Suppose newspaper reporters brainstormed some headlines for an article on this experiment. These are their suggested headlines:

A. "New Treatment Helps Pericarditis Patients"

B. "Colchicine Tends to Improve Treatment for Pericarditis"

C. "Pericarditis Patients May Get Help"

7. Which of the headlines above would be best to use for the article? Explain why.

Exercises 8–10

What you should look for when evaluating an experiment:

- Were the subjects randomly assigned to treatment groups?
- Was there a control group or a comparison group?
- Were the sample sizes reasonably large?
- Do the results show a cause-and-effect relationship?

Read the abstracts of the two articles below. Write a few sentences evaluating these articles using the guidelines above.

8. The study "Semantic Memory Functional MRI and Cognitive Function After Exercise Intervention in Mild Cognitive Impairment" (*Journal of Alzheimer's Disease*, August 2013) was performed to see if exercise would increase memory retrieval in older adults with mild cognitive impairment (associated with early memory loss).

    Website: http://iospress.metapress.com/content/xm8t241628h37h7t/

9. The article "Effects of Bracing in Adolescents with Idiopathic Scoliosis" (*New England Journal of Medicine*, October 2013) reports on the role of bracing patients with adolescent idiopathic scoliosis (curvature of the spine) for prevention of back surgery.

    Website: www.nejm.org/doi/full/10.1056/NEJMoa1307337

10. View the report by Tom Bemis (Market Watch, *Wall Street Journal*, August 13, 2013) about the type of car driven by a person and the person's driving behavior.

    Website: http://live.wsj.com/video/bmw-drivers-really-are-jerks-studies-find/29285015-BB1A-4E41-B0C0-0A41CB990F60.html#!29285015-BB1A-4E41-B0C0-0A41CB990F60

    Is the title "BMW Drivers Really Are Jerks" an accurate title for these reported studies? Why or why not? If not, suggest a better title.

> **Lesson Summary**
>
> - A cause-and-effect relationship can only be shown by a well-designed experiment.
>
> - Randomly assigning the subjects to treatment groups evens out the effects of extraneous variables to create comparable treatment groups.
>
> - A control group (which may be a placebo group) or a comparison group (a standard treatment) is sometimes included in an experiment so that you can evaluate the effect of the treatment.
>
> - The number of subjects in each treatment group (sample size) should be large enough for the random assignment to experimental groups to create groups with comparable variability between the subjects.

Problem Set

Read the following articles and summaries. Write a few sentences evaluating each one using the guidelines given in the lesson.

1. The article "Emerging Technology" (*Discover Magazine*, November 2005) reports a study on the effect of "infomania" on IQ scores.

   Website: discovermagazine.com/2005/nov/emerging-technology

2. In *The New England Journal of Medicine*, October 2013, the article "Increased Survival in Pancreatic Cancer with nab-Paclitaxel Plus Gemcitabine" reports on an experiment to test which treatment, nab-paclitaxel plus gemcitabine or gemcitabine alone, is the most effective in treating advanced pancreatic cancer.

   Website: www.nejm.org/doi/full/10.1056/NEJMoa1304369

3. Doctors conducted a randomized trial of hypothermia in infants with a gestational age of at least 36 weeks who were admitted to the hospital at or before six hours of age with either severe acidosis or perinatal complications and resuscitation at birth and who had moderate or severe encephalopathy. The trial "Whole-Body Hypothermia for Neonates with Hypoxic–Ischemic Encephalopathy" tested two treatments, standard care and whole-body cooling, for 72 hours.

   Website: www.nejm.org/doi/full/10.1056/NEJMcps050929

This page intentionally left blank

This page intentionally left blank

This page intentionally left blank

This page intentionally left blank